A DICTIONARY
OF NAMED EFFECTS
AND LAWS

A DICTIONARY
OF NAMED EFFECTS
AND LAWS

IN CHEMISTRY, PHYSICS AND MATHEMATICS

D. W. G. BALLENTYNE
B.Sc. Ph.D. F.Inst.P.

D. R. LOVETT
B.Sc. A.R.C.S. D.I.C. Ph.D. M.Inst.P.

LONDON

CHAPMAN AND HALL

First published 1958
by Chapman and Hall Ltd.
11 New Fetter Lane London EC4P 4EE
Second edition 1961
Third edition 1970
Reprinted 1972

First issued as a Science Paperback 1972
Reprinted 1976

© *1958, 1961 D.W.G.Ballentyne and L.E.Q.Walker*
© *1970 D.W.G.Ballentyne and D.R.Lovett*

Set by DDP Print & Design Co Ltd, Colchester
Printed in Great Britain by
Whitstable Litho Ltd, Whitstable, Kent

ISBN 0 412 20970 5

Distributed in the U.S.A. by Halsted Press, a Division
of John Wiley & Sons, Inc., New York

PREFACE TO THE THIRD EDITION

The third edition of this book has been completely revised, and some three hundred new entries have been added. This work has been undertaken by a changed team of authors as Mr. L. E. Q. Walker has retired from active participation and has now been replaced by Dr. D. R. Lovett, a member of the Physics Department of the University of Essex.

Two major alterations to the format of the previous volume have been made. Firstly, symbols which occur a number of times are defined once and for all in a table after the Preface. Secondly, the entries have in some cases been rearranged so that they are now in strict alphabetical order.

The Appendix of units has been greatly extended to cover, it is hoped, all named units at present used by scientists in these fields. However, a book of this kind must by its very nature contain some omissions and the authors would be grateful for communications from any readers who come across deficiencies of this type.

PREFACE TO THE FIRST EDITION

Every science has its own vocabulary. It is impossible to read many pages of any scientific book without encountering words which possess a specific and unique meaning to the particular scientific subject with which the book deals. Some of these words are proper nouns, either used substantively or, more rarely, adjectivally. These are the names of scientists who have investigated a particular phenomenon or who have discovered some scientific law or relation or who have worked in some field with which their name has become historically connected.

It is with such names that this book is concerned. It is by no means intended to be read as a text-book but rather to be consulted as a dictionary whenever the reader, possibly an expert in one branch of science, is confronted by a mention of a relation or rule or law of someone or other who worked, maybe, in quite another field. He may not feel inclined to delve very deeply into the origins of the phrase. He may, in fact, wish to obtain such information as may enable him to proceed, as quickly as possible, with his reading. It is partly in an endeavour to help him that this glossary has been compiled.

Classification by subject matter has not been attempted and entries appear in alphabetical order.

SYMBOLS

(Unless othwise stated; further symbols are defined within the entries.)

c	velocity of light
e	electron charge
F	Faraday constant
g	acceleration due to gravity
h	Planck's constant
$\hbar =$	$h/2\pi$
k	Boltzmann's constant
m	electron mass
N	number of molecules
p	pressure
R	gas constant (per mole)
t	time
T	temperature (absolute scale)
V	volume

exp	exponential
$j =$	$\sqrt{-1}$
ln	logarithm to base e
log	logarithm to base 10
O . . .	term of order . . .

A

Abbe's Sine Condition
When the pencil of rays forming an optical image is of finite aperture, the condition that a magnification of the image can be obtained which is independent of the zone of the lens traversed is given by

$$\frac{n \sin a}{n' \sin a'} = \text{constant} = m$$

where n and n' are the refractive indices of object and image space, a and a' are the angles with which a ray leaves and (after refraction) reaches the axis again respectively and m is the lateral magnification.

Abbe's Theory (of the Diffraction of Microscopic Vision)
For the production of a truthful image of an illuminated structure by a lens it is necessary that the aperture be wide enough to transmit the whole of the diffraction pattern produced by the structure. If only a portion of the diffraction pattern is transmitted the image will correspond to an object whose diffraction pattern is identical with the portion passed by the lens. If the structure is so fine or the lens aperture so narrow that none of the diffraction pattern is transmitted the structure will be invisible regardless of the magnification.

Abderhalden Reaction
To detect the presence of protective ferments in the blood, any change of optical constants is noted when the serum is mixed with albumen. The test can be used to detect pregnancy and certain malignant diseases.

Abegg's Rule
The sum of the maximum positive valency exhibited by an element and its maximum negative valency equals 8. This rule is true generally for the elements of the 4th, 5th, 6th and 7th groups of the Periodic Table.

Abelian Group
If a group of elements A, B, C, . . . finite or infinite in number have the property that AB = BA (commutative property) for every element, then the group is known as an Abelian Group.

Abel's Identities
(1) If $A_s = a_n + a_{n+1} + \ldots a_s$ then

$$\sum_{s=p}^{m} a_s b_s = \sum_{s=p}^{m} (b_s - b_{s+1}) A_s - b_p A_{p-1} + b_{m+1} A_m$$

(2) If $A_s' = a_s + a_{s+1} + a_{s+2} + \ldots$ then

$$\sum_{s=p}^{m} a_s b_s = \sum_{s=p}^{m-1} (b_{s+1} - b_s) A_{s+1}' + b_p A_p' - b_m A_m' + 1$$

Abel's Inequality
If $u_n \to 0$ monotonically for integer values of n, then $| \sum_{n=1}^{m} a_n u_n | \leqslant A u_1$ where A is the greatest of the sums $|a_1|, |a_1 + a_2|, |a_1 + a_2 + a_3| \ldots |a_1 + a_2 \ldots + a_m|$.

Abel's Integral Equation
A particular type of **Volterra integral** of the form

$$\int_a^z \frac{\chi(z_0)}{(z - z_0)^a} \, dz_0 = \phi(z) \qquad (0 < a < 1).$$

The solution of the equation is

$$\chi(z) = \frac{\sin(\pi a)}{a} \left[\frac{\varphi(a)}{(z-a)^{1-a}} + \int_a^z \frac{\phi(z_0) \, dz_0}{(z - z_0)^{1-a}} \right]$$

The equation, with $a = \frac{1}{2}$, has particular application to the time of fall of a particle along a smooth curve in the vertical plane when $\phi(z) = $ time.

Abel's Test (for Convergence)
If Σu_n converges and $a_1, a_2, a_3, \ldots$ is a decreasing sequence of positive terms, then $\Sigma a_n u_n$ is convergent.

Abel's Test (for Infinite Integrals)
If $\int_a^\infty f(x) \, dx$ converges, and if for every value of y such that $b \leqslant y \leqslant c$ the function $g(x, y)$ is neither negative nor increasing with x, then $\int_0^\infty f(x) g(x, y) dx$ is uniformly convergent with respect to y in the range $b \leqslant y \leqslant c$.

2

Abel's Theorem (on Multiplication of Series)

If $c_n = a_0 b_n + a_1 b_{n-1} + \ldots + a_n b_0$ then the convergence of

$\sum\limits_{n=0}^{\infty} a_n, \sum\limits_{n=0}^{\infty} b_n$ and $\sum\limits_{n=0}^{\infty} c_n$ is a sufficient condition that

$$\left(\sum\limits_{n=0}^{\infty} a_n \right) \left(\sum\limits_{n=0}^{\infty} b_n \right) = \sum\limits_{n=0}^{\infty} c_n.$$

Abney Law

If the colour of a spectral line is desaturated by the addition of white light, its colour to the eye shifts towards the red if its wavelength is less than 5700Å and to the blue if its wavelength is greater than this value.

Acree's Reaction

As a test for proteins, when formaldehyde solution containing a trace of ferric chloride is added to the protein solution and concentrated sulphuric acid introduced below the mixed solution, a violet ring is formed.

Adams-Bashforth Process

A method of numerically integrating ordinary differential equations. Starting with Gregory's backwards formula (*see* **Gregory's Interpolation Formulae**), $f(x)$ is expanded and integrated to give

$$\frac{1}{h} \int_{x_0}^{x_0+h} f(x)dx = f(x_0) + \left(\tfrac{1}{2}\nabla + \frac{5}{12}\nabla^2 + \frac{3}{8}\nabla^3 + \frac{251}{720}\nabla^4 + \frac{95}{288}\nabla^5 + \ldots \right)f($$

Hence, if $dy/dx = f(x, y)$ and y and f are known for values of x up to $x = x_0$, a value for y corresponding to $x = x_0 + h$ can be obtained from f at x_0 and the backwards differences.

Agnesi—Witch of

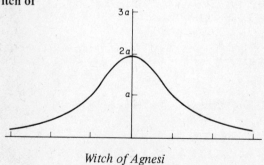

Witch of Agnesi

A curve whose equation is $x^2 y = 4a^2(2a - y)$

Airy's Disc

The diffraction pattern formed by plane light waves from a point source passing through a circular aperture consists of a bright central disc, known as Airy's disc, surrounded by further rings. Airy in 1934 obtained the intensity distribution across the pattern in terms of **Bessel Functions** of order unity. The radius of the central disc is given by

$$\frac{0\cdot61\,\lambda}{n\,\sin\,U}$$

where λ is the wavelength *in vacuo*, U is the semi-angle of the emergent cone of light from the aperture and n is the refractive index on the image side of the aperture. The form of the diffraction pattern has particular application to the calculation of the resolving power of telescopes and other optical instruments.

Airy's Equation

An equation for multiple beam interference for light transmitted through a plane parallel plate. The intensity of the transmitted light is given by

$$I = \frac{I_0\,T^2}{(1-R)^2\left[1+\dfrac{4R\,\sin^2\,\delta/2}{(1-R)^2}\right]}$$

where I_0 is the intensity of the incident beam, R is the reflectivity of the surfaces of the plate and T their transmissivity, and δ is the phase difference between the directly transmitted light and light which is once reflected from the two internal surfaces of the plate. (*See also* **Fabry-Perot Fringes**.)

Airy's Integral

The integral solution of the differential equation

$$\frac{d^2y}{dz^2} - zy = 0$$

defined as

$$A\mathrm{i}(z) = \frac{1}{2\pi\mathrm{j}}\int_L e^{\,tz-t^3/3}\,dt = \frac{1}{\pi}\int_0^\infty \cos\,(sz+s^3/3)\,ds$$

4

where L is a contour as shown in the diagram.

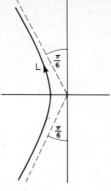

Airy Integral

Airy's Points

The optimum points at which a bar must be suspended horizontally to make bending a minimum. The distance apart of the points is

$$\frac{l}{\sqrt{(n^2 - 1)}}$$

where l is the length of the bar and n the number of supports.

Aitken's Formula

If a series of numbers $u_1, u_2, \ldots u_n, u_{n+1}, u_{n+2}, \ldots$, is expected to converge slowly to an unknown limit u, then u can be estimated by assuming

$$u_n = u - e; u_{n+1} = u - ke; u_{n+2} = u - k^2 e \ (k < 1)$$

whence, by eliminating k and e;

$$u = \frac{u_{n+2}\,u_n - u_{n+1}^2}{u_{n+2} - 2u_{n+1} + u_n} = u_{n+2} - \frac{(u_{n+2} - u_{n+1})^2}{u_{n+2} - 2u_{n+1} + u_n}$$

Almén-Nylander Test

A test for the presence of sugar, consisting in the reduction of a bismuth salt solution to metallic bismuth.

5

Alvén Waves

In 1942 Alvén predicted the possible existence of magnetohydrodynamic waves—in the simplest case transverse oscillations of magnetic field lines carrying with them a surrounding non-viscous perfectly conducting fluid. The velocity of propagation is given by

$$v^2 = \frac{\mu H^2}{4\pi\rho}$$

where H is the magnetic field intensity, ρ the density of the fluid and μ the magnetic permeability.

Amagat *See* Appendix
Amagat-Leduc Rule

According to E. H. Amagat and A. Leduc the volume occupied by a mixture of gases is equal to the sum of the volumes that the constituent gases would individually occupy at the temperature and pressure of the mixture. The Amagat-Leduc rule and **Dalton's Law of Partial Pressures** are identical for the perfect gas.

Amonton's Law

In any gas, whose volume and mass are kept constant, the same rise in temperature produces the same increase of pressure.

Ampere *See* Appendix

Ampere-turn *See* Gilbert (Appendix)

Ampère's Law

The magnetic force produced at a point P by a current flowing in a conductor is given by

$$\int \frac{I \, ds \, \sin\theta}{r^2}$$

where I is the current flowing in an element of the circuit ds, r is the length of the line joining ds and the point P, and θ is the angle between ds and r. The law is also called **Laplace's Law** or the **Biot Savart Relation**.

Alternatively, Ampère's Law is sometimes stated as

$$\oint B \cdot ds = 4 \pi I$$

where the integral is over a closed path enclosing a total current I which sets up magnetic induction B.

Andrade's Creep Law
Andrade showed that when a load is applied at the beginning of a creep test the instantaneous elastic elongation is followed by a transient state in which strain varies as $t^{1/3}$ and finally a steady state is reached in which there is a constant rate of creep under constant effective stress.

Ångström *See* **Appendix**

Ångström's Formula
For the scattering effect of dust in the atmosphere

$$S = A\lambda^{-B}$$

where λ is the wavelength, B depends on the particle size and A is Ångström's coefficient.

Angus-Smith Process
When iron is heated to about 600°F and then immersed in a solution of coal-tar in oil and paranaphthalene an anti-corrosive layer is formed. *See* **Bower-Barff Process.**

Antoine Equation
An expression for the vapour pressure p of a condensed solid or liquid as a function of absolute temperature T;

$$\ln p = A - \frac{B}{C + T}$$

where A, B and C are constants obtained experimentally.

Antonoff's Rule
The interfacial surface tension γ_{AB} between two saturated liquid layers, A and B, in equilibrium, is equal to the difference of the surface

7

tensions against vapour or air of the two mutually saturated solutions, i.e.

$$\gamma_{AB} = \gamma_A - \gamma_B$$

Apjohn's Formula

The pressure of water vapour in the air can be calculated from the temperatures of the wet and dry bulbs of a hygrometer (t_w and t respectively) by the formula

$$p_t = p_w - 0 \cdot 00075 H(t - t_w)\{1 - 0 \cdot 008(t - t_w)\}$$

where p_w is the saturated vapour pressure at temperature t_w and H is the barometric height.

Apollonius' Circle

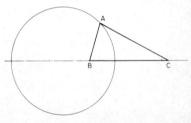

The locus of the vertex, A, of a triangle of given base BC such that the sides AB and AC are in a given ratio $\lambda:1$ is a circle with as diameter the line joining the points which divide the base BC internally and externally in the ratio $\lambda:1$.

Apollonius' Theorem

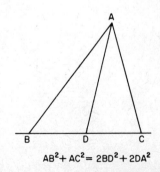

$$AB^2 + AC^2 = 2BD^2 + 2DA^2$$

8

In any triangle, the sum of the squares on two sides is equal to twice the square on half the base together with twice the square on the median drawn to the base.

Arbusov Reaction (or Rearrangement)

In the Arbusov rearrangement, an alkyl phosphite is heated with a small amount of the corresponding alkyl halide. Initially a phosphonium salt is formed which decomposes to give a dialkyl alkyl phosphonate. The formation of the phosphonium salt is so fast that if an equivalent amount of a different alkyl halide is used, up to 95% yield of a product having the new alkyl group attached to the phosphorus can be obtained.

$$(RO)_3P + R'X \rightarrow (RO)_3P^+ \overset{-}{-}X \rightarrow (RO)_2P\!-\!O + RX$$
$$\underset{R'}{|} \qquad\qquad \underset{R'}{|}$$

Archimedes' Axiom

If a and b are any two positive rationals, an integer n exists, such that $nb > a$.

Archimedes' Principle

When a body is immersed in a fluid there is an apparent loss in weight which is equal to the weight of fluid displaced.

Archimedes' Spiral

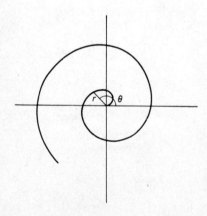

A point, moving uniformly along a line which rotates uniformly about a fixed point describes a spiral of Archimedes. The equation is: $r = a\theta$.

Argand Diagram

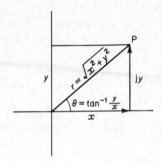

The representation of a complex quantity

$$z = x + jy$$

where x is the real and y the imaginary part, by a point P whose Cartesian coordinates are x, y. If z is expressed in De Moivre's form

$$z = re^{j\theta} = r(\cos\theta + j\sin\theta)$$

$r = \sqrt{(x^2 + y^2)}$, the length of the line from the origin O to P, and $\theta = \tan^{-1} y/x$ where θ is the angle between OP and the x axis.

Armstrong-Baeyer Benzene Formula *See* **Kekulé Benzene Formula**

Arndt-Eistert Synthesis
An aliphatic or aromatic acid may be converted to the next higher homologue by means of diazomethane. The acid chloride is treated with two moles of diazomethane and the diazoketone is warmed with water in the presence of a silver catalyst.

$$R.COCl + 2CH_2N_2 \rightarrow R.CO.CHN_2 + CH_3Cl + N_2.$$

$$RCOCHN_2 + H_2O \xrightarrow[50-80\%]{Ag} RCH_2COOH + N_2$$

10

Arnold's Test
A test for diacetic acid in urine. When treated with solutions containing
p-amino acetophenone and sodium nitrite a purple or violet colour is
produced.

Arrhenius' Equation (for Reaction Velocity)
The most satisfactory method of expressing the effect of temperature
on reaction velocity is due to S. Arrhenius and may be expressed by
the equation

$$k = se^{-E/RT}$$

where k is the specific reaction rate, E is the activation energy of the
reaction, and s is the frequency factor.

Arrhenius' Theory
In 1887, S. Arrhenius evolved a theory of electrolytic dissociation on
the basis of calculations of i, \the **Van't Hoff Factor**, from electromotive
force measurements and conductivity measurements. When an acid,
base or salt is dissolved in a polar solvent a certain proportion of the
molecules become spontaneously ionized:

$$AB \rightleftharpoons A^+ + B^-$$

The degree of dissociation, a, was shown to vary with the concentration
and to approach unity at infinite dilution. Electrolysis can then be
easily explained as the independent migration of the negative and
positive ions to the oppositely charged electrode. For the modern views
on electrolytic dissociation see the **Onsager Conductivity Equation**.

Aston Dark Space
In a gas discharge the dark space in the immediate vicinity of the
cathode, in which the emitted electrons have a velocity insufficient to
excite the gas.

Aston Rule
The atomic weights of isotopes are approximately integers, and
deviations of the atomic weights of the elements from integers are due
to the presence of several isotopes with differing weights.

Atkinson Cycle

A working cycle for an internal-combustion engine where the expansion ratio exceeds the compression ratio. It is more efficient theoretically than the Otto cycle but is difficult to achieve in practice.

Auger Effect

When a K electron is ejected from an atom, an outer electron can fall into the K shell and give up its energy in two ways:

I. By radiation which gives rise to x-ray emission lines.

II. By transferring its energy to one of the outer electrons which is ejected from the atom. This is known as the Auger Effect.

Austin-Cohen Law

$$E = \frac{377Ih}{\lambda r}\, e^{-ar}$$

where E is the electric field strength from an aerial of effective height h and carrying a uniform current I; r is the distance between transmitter and receiver; λ is the wavelength; and a is a constant depending upon the terrain over which the wave passes.

Avogadro's Law

Equal volumes of all gases, under the same conditions of temperature and pressure, contain equal numbers of molecules.

Avogadro's Number

The number of individual atoms in a gramme-atom, of ions in a gramme-ion, or of molecules in a gramme-molecule, is called the Avogadro number. The accepted value is $6 \cdot 023 \times 10^{23}$ in chemistry where it is the number of atoms in 16 grammes of natural oxygen O or $6 \cdot 02497 \times 10^{23}$ in physics where it is the number of atoms in 16 grammes of oxygen isotope O^{16}.

Azbel-Kaner Resonance

A type of cyclotron resonance in high purity metals at liquid helium temperatures, first observed by Azbel and Kaner in 1957. A magnetic field is applied parallel to the surface of the metal in the same direction as an oscillating electric field. Current can only flow in a region next to the surface having a depth called the 'anomalous skin depth'. The radius of the electron orbits under the above conditions is much larger than this skin depth and so electrons can only be accelerated once per cycle and only if their orbits enter the skin depth region. The frequency of the electric field must be the cyclotron frequency or an integral multiple of it.

B

Babinet's Principle
The diffraction patterns obtained from a complementary pair of diffraction screens, that is two screens in which the opaque parts of one are replaced by transparent parts in the other, are the same.

Babo's Law
The lowering of the vapour pressure of a solvent by the addition of a non-volatile solute is proportional to its concentration.

Back-Goudsmit Effect *See* Paschen-Back Effect

Badger Rule
An empirical relationship between the force constants and the vibrational frequencies of electron orbits in diatomic molecules.

Baeyer Strain Theory
The angle subtended by the corners and centre of a regular tetrahedron (109° 28′) lies between the values for the angle of a regular pentagon (108°) and a regular hexagon (120°). According to Baeyer

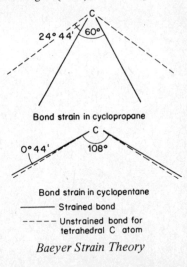

Bond strain in cyclopropane

Bond strain in cyclopentane

——— Strained bond

- - - - Unstrained bond for tetrahedral C atom

Baeyer Strain Theory

14

therefore *cyclo*paraffins in which the carbon atoms are co-planar are only possible when the minimum strain is set up between the carbon bonds and for this reason the 5- and 6-membered rings should be most stable, as is found in practice.

The carbon bonds must be strained together by $+24°44'$ in order to form *cyclo*propane and therefore this compound should be very difficult to prepare and unstable, whereas in the case of *cyclo*pentane the strain is only $+0°44'$, hence this compound should be stable. Strain when the bonds are forced together is designated as positive whereas when they are forced apart the sign is negative. This theory has been modified by a number of workers. *See* **Sachse-Mohr Theory.**

Baeyer-Villiger Reaction
Ketones when oxidized with peroxy acids yield esters. The preferred reagent is trifluoro peroxy acetic acid or mono peroxy maleic acid.

Baily's Beads
Just before a solar eclipse becomes total the advancing dark limb of the moon appears to break up into a series of bright points.

Baker-Nathan Effect (Hyperconjugation)
According to these workers the methyl group in propylene and certain other compounds is more electron releasing than would be normally expected by the inductive effect. The phenomenon is only exhibited when the methyl group is attached to an unsaturated carbon atom.

Balmer *See* Appendix

Balmer Series
A series of lines observed in the atomic spectrum of hydrogen. The wave number of the lines of a spectral series may be represented by the equation

$$\bar{\nu} = R \left(\frac{1}{n_1^2} - \frac{1}{n_2^2} \right)$$

where R is the Rydberg constant. Various values of n_1 and n_2 give a number of spectral series:

Series	n_1	n_2
Lyman	1	2, 3, 4, ...
Balmer	2	3, 4, 5, ...
Paschen	3	4, 5, 6, ...
Brackett	4	5, 6, 7, ...
Pfund	5	6, 7, 8, ...

(*See also* **Ritz Combination Principle** and **Bohr Theory of Spectral Emission.**)

Balz-Schiemann Reaction

Fluorobenzenes may be prepared by this reaction. When borofluoric acid is added to a diazonium salt the insoluble diazonium borofluoride is precipitated, collected by filtration, dried and heated gently when it is decomposed into fluorobenzene, nitrogen and boron trifluoride.

$$C_6H_5N_2Cl + HBF_4 \rightarrow C_6H_5N_2BF_4 + HCl$$
$$\downarrow$$
$$C_6H_5F + N_2 + BF_3$$

Bamberger's Formula

A structural formula for naphthalene in which the valencies of the benzene rings point towards the centres. The formula does not involve double bonds as in the Kekulé formula and bears no relation to the real structure of naphthalene which is a resonance hybrid.

Bardeen-Cooper-Schrieffer (BCS) Theory of Superconductivity

A quantum theory of superconductivity in which interaction between pairs of electrons gives for the electron system a ground state which is separated from the excited states by an energy gap. The presence of this energy gap accounts for the normal/superconducting phase transition and the critical magnetic field which leads to the destruction of superconductivity. The theory accounts for the **London Penetration Depth** and the **Pippard Coherence Length.**
(*Phys. Rev.* (1957), **106**, 162 and **108**, 1175)

Barfold's Test

A test for monosaccharoses relying on the reduction of cupric acetate to red cuprous oxide.

Barkhausen Effect
When a ferromagnetic material is magnetized in an increasing magnetic field, the magnetization does not increase smoothly with field. This is because of the irregular fluctuations in the motion of the domain walls which separate regions of differing directions of magnetization in the material.

Barkhausen-Kurtz Oscillations
Oscillations occurring in an electronic valve, wherein electrons of high velocity are expelled from the filament, meet the repelling effect of the grid-anode field, stop, return through the grid, and so on, eventually landing on the grid wires. The current accompanying this motion may be used to produce oscillations of frequencies between 300 and 1500 MHz.

Barlow's Rule
The volume occupied by the various atoms in a given molecule is proportional to the valencies of the atoms. With multivalent elements the lowest value of valency is used.

Barnett Effect
The magnetization acquired by an initially non-magnetized and stationary body when the body is rotated. This rotation will produce the same intensity of magnetization in the body as a uniform axial field of magnetic intensity $H = \gamma\Omega$ where Ω is the angular velocity of rotation and γ the gyromagnetic ratio (the ratio of the angular momentum to the magnetic moment) of the elementary carrier. The effect has been measured in ferromagnetic materials.

The converse effect, the rotation of a body of magnetic material when a magnetic field is applied parallel to its axis, is called the **Einstein de Haas Effect** or the **Richardson Effect**.

Bartlett Force
An exchange force between nucleons in which spin (equivalent to charge plus position) is exchanged. (*See also* **Majorana Force**.)

Bart Reaction
When a diazonium salt is treated with sodium arsenite in the presence of copper the diazonium group is replaced by an arsonic group.

17

$$C_6H_5N_2Cl + Na_3AsO_3 \xrightarrow{CuSO_4} C_6H_5AsO_3Na_2 + N_2 + NaCl$$
$$\downarrow HCl$$
$$C_6H_5AsO_3H_2$$

Bateman Equations

Equations applicable to a chain of radioactive nuclides denoted by subscripts 1 to n. The nth nuclide decays into the $(n + 1)^{th}$ nuclide with a disintegration constant λ_n, and at any time t there are N_n atoms of the nth nuclide. Then,

$$\frac{dN_1}{dt} = -\lambda_1 N_1$$

$$\text{and } \frac{dN_n}{dt} = \lambda_{n-1} N_{n-1} - \lambda_n N_n.$$

If at $t = 0$ only the parent substance is present, and $N_1 = N_1^0, N_2^0 = N_3^0 = \ldots = N_n^0 = 0$, then the number of atoms of the nth member of the chain is

$$N_n(t) = \sum_1^n \frac{\lambda_1 \lambda_2 \ldots \lambda_{n-1} N_1^0 \, e^{-\lambda_n t}}{(\lambda_1 - \lambda_n)(\lambda_2 - \lambda_n) \ldots (\lambda_{n-1} - \lambda_n)}.$$

Baudisch Reaction

o-nitrosophenol may be prepared from benzene, by the action of a nitrosyl radical in the presence of an oxidizing agent. The nitroso-radical may be prepared by the reduction of nitrous acid or the oxidation of hydroxylamine in the presence of a copper salt which both stabilizes the free radical and prevents the formation of p-nitrosophenol.

Baumé Scale

A hydrometer scale in which $0°$ represents the specific gravity of water at $12.5°C$ and $10°$ the specific gravity of a 10% solution of sodium chloride at the same temperature. Also known as the **Lunge Scale**.

Bauschinger Effect
If, at a temperature at which work-hardening occurs, a metal is
deformed plastically under tension and then unloaded, the mechanical
properties of the metal for further tensile loading are found to be
different to those for compressive loading. The magnitude of the effect
usually depends on the size of the tensile stress applied before the
compressive loading.

Beattie-Bridgeman Equation of State
For a real gas,

$$p = \frac{RT(1 - \epsilon)}{V^2} \cdot (V + B) - \frac{A}{V^2}$$

where p is the pressure of the gas, V is the volume,

$$A = A_0 \left(1 - \frac{a}{V}\right) \quad B = B_0 \left(1 - \frac{b}{V}\right) \quad \epsilon = \frac{c}{VT^3}$$

This equation with five adjustable constants can be made to represent
the behaviour of a real gas above the triple point.

Becke Line
When two adjacent substances, for instance a crystal immersed in a
liquid, are viewed under a microscope, then, if their refractive indices
are close in magnitude, a narrow but bright line of light appears at
the junction. The line is called the Becke line and arises from the
combination of reflected and refracted light rays. If it is the crystal
which has the higher refractive index, then as the objective of the
microscope is raised, the Becke line appears to move into the crystal.

Beckmann Rearrangement
When treated with certain reagents, such as sulphuric acid, hydrochloric
acid or phosphorus pentachloride, ketoximes undergo rearrangement
to the acid amide. As this reaction is specific for anti-ketoximes it may
be used to determine the configuration.

Becquerel Effect

If two similar electrodes are immersed in an electrolyte, a current flows when one of the electrodes is illuminated.

Beer's Law

If I is the intensity of the light transmitted through a solution at a given wavelength and I_0 is the intensity of the incident light, then, according to Beer,

$$I = I_0 \; 10^{-\mu cd}$$

where c is the concentration of solution in gramme-molecules per litre and d is the thickness of solution through which the light is transmitted, μ is known as the molar extinction coefficient of the solute for the particular wavelength. The Law is sometimes used in the form

$$I = I_0 \exp(-acd)$$

where a is the molar absorption coefficient.

Beilby Layer

A microcrystalline or amorphous layer formed on the surface of metals by polishing.

Beilstein's Test

This test detects the presence of halogens in many organic compounds. The organic halogen compound is heated in contact with cupric oxide (usually in the form of heated copper gauze) in a Bunsen flame. The volatile copper halide which is formed colours the flame green or blue. The test is not specific as certain compounds (derivatives of pyridine, quinoline and purines for example) also give a positive result due to the formation of copper cyanide.

Bel *See* Appendix

Benedict's Solution

A modification of **Fehling's Solution** used to detect reducing sugars. 86·5gm of sodium citrate duodecahydrate and 50gm of anhydrous sodium carbonate are dissolved in 350c.c. 8·65gm of hydrated copper sulphate dissolved in 50c.c. of water are added with constant stirring.

20

The solution is diluted to 500c.c. To 5c.c. of Benedict's Solution add 0·4c.c. of 2% solution of the carbohydrate, warm and allow to cool to room temperature. The presence of a reducing sugar is indicated by the precipitation of cuprous chloride.

Bergmann Method (of Polypeptide Synthesis)

During the preparation of polypeptides of known structure it is necessary to control the position in which the condensation of a second amino acid can occur and in general to prevent ring closure. Many methods have been advanced to block the amino group of an amino acid during the preparation of a polypeptide, but that due to Bergmann is of must utility. The blocking agent in this case is benzyl oxycarbonylchloride, $C_6H_5CH_2O.CO.Cl$. The synthesis may be exemplified by

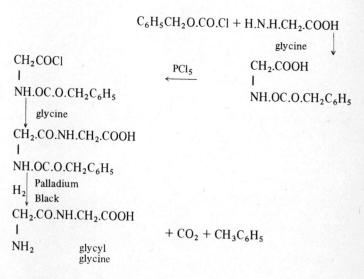

Bernoulli-Euler Law

In a homogeneous bar, the curvature of the central fibre is proportional to the bending movement.

Bernoulli, Lemniscate of

A curve whose equation is

$$(x^2 + y^2)^2 - a^2(x^2 - y^2) = 0$$

21

Bernoulli Polynomials

Bernoulli polynomials $B_r{}^{(n)}(x)$ of order n are given by the identity

$$\frac{t^n e^{xt}}{(e^t - 1)^n} = \sum_{r=0}^{\infty} \frac{t^r}{r!} B_r{}^{(n)}(x)$$

where $n = 0, 1, 2 \ldots$. In addition the polynomials $B_r{}^{(n)}(0) = B_r{}^n$ are called Bernoulli numbers of order n except for $n = 1$ when they are usually simply called **Bernoulli numbers**, B_r.

Bernoulli's Differential Equation

Bernoulli's differential equation is of the form

$$\frac{dy}{dx} + f(x)\, y = g(x)\, y^n$$

This equation may be made linear by the substitution

$$y = u^{1/(1-n)};$$

Bernoulli's Equation (of Liquid Motion)

Liquid escapes through an orifice of cross-sectional area a with a velocity v given by

$$v^2 = 2gh\, \frac{A^2}{A^2 - a^2}$$

where A is the surface area and h is the head of liquid and g is the acceleration due to gravity. For large values of A this equation reduces to **Torricelli's Law**.

Bernoulli's Inequality

If $x > 0, y > 0$ and $n \neq 0$ or 1, then
(1) $x^n - 1 > n(x - 1)$ if $x \neq 1$
(2) $nx^{n-1}(x - y) > x^n - y^n > ny^{n-1}(x - y)$ if $x \neq y$
unless $0 < n < 1$ when the inequality sign is reversed in both cases.

Bernoulli's Theorem

A sequence of independent trials in which the probability of 'success' is the same for each is termed a series of **Bernoulli trials**. If s is the number of 'successes' in a series of n independent trials and p is the probability of an individual 'success', then Bernoulli's theorem states that the probability P of the relative frequency of 'successes' ($= s/n$) differing from p by more than ϵ tends to zero as $n \to \infty$ for any $\epsilon > 0$; i.e.

$$\lim_{n \to \infty} P\left\{ \left| \frac{s}{n} - p \right| > \epsilon \right\} = 0 \quad (\epsilon > 0)$$

Bernoulli's Theorem in a Field of Flow
At every point in a steadily flowing fluid the sum of the pressure head, the velocity head and the height is constant.

Berthelot Equation
The Berthelot Equation is a modified gas equation for real gases of the form

$$PV = RT \left\{ 1 + \frac{9}{128} \cdot \frac{PT_c}{P_c T} \left(1 - 6 \frac{T_c^2}{T^2} \right) \right\}$$

where P_c and T_c are the critical pressure and critical temperature of the gas. The equation holds best at low pressures and is applicable only over a limited range.

Berthelot-Thomsen Principle
In 1867, M. Berthelot postulated that, of all possible chemical reactions, the one accompanied by the greatest development of heat will occur. Changes of state must be excluded from this rule and there are other obvious deviations.

Bertrand's Test (for Convergence)

If $f(x) = O(x^{\lambda-1})$ or if $f(x) = O(x^{-1} [\ln x]^{\lambda-1})$

then $\displaystyle\int_a^\infty f(x)\,dx$ converges when $\lambda < 0$.

Bessel-Clifford Differential Equation

$$u \frac{d^2 y}{du^2} + (n+1) \frac{dy}{du} + y = 0$$

23

reduces to Bessel's equation (*see* **Bessel Functions**) with the substitution

$$y = z\, u^{-n/2} \text{ and } u = x^2/4.$$

The equation has a solution of the form

$$y = u^{-n/2}\, J_n\, (2\sqrt{u}) + u^{-n/2}\, N_n\, (2\sqrt{u}).$$

Bessel Functions

Bessel's equation

$$\frac{d^2z}{dx^2} + \frac{1}{x}\frac{dz}{dx} + \left(1 - \frac{n^2}{x^2}\right)z = 0$$

has as its general solution

$$z = A J_n\, (x) + B N_n\, (x)$$

where

$$J_n\, (x) = \frac{x^n}{2^n\Gamma(n+1)} \left\{ 1 - \frac{x^2}{2\,(n+1)} + \frac{x^4}{2^4 2!\, (n+1)\,(n+2)} \right.$$

$$\left. - \frac{x^6}{2^6 3!\, (n+1)\,(n+2)\,(n+3)} + \cdots \right\}$$

is a Bessel Function of the nth order or a Cylindrical Harmonic, and

$$N_n\, (x) = \frac{2}{\pi}\, J_n\, (x) \left\{ \ln\frac{x}{2} + \gamma \right\} - \frac{1}{\pi} \left(\frac{x}{2}\right)^{-n} \sum_{k=0}^{k=n-1} \frac{(n-k-1)!}{k!} \left(\frac{x}{2}\right)^{2k}$$

$$- \frac{1}{\pi} \left(\frac{x}{2}\right)^n \sum_{k=0}^{k=\infty} \frac{(-1)^k}{(n+k)!\,k!} \left(1 + \tfrac{1}{2} + \tfrac{1}{3} + \cdots \right.$$

$$\left. \cdots + \frac{1}{k} + 1 + \tfrac{1}{2} + \tfrac{1}{3} + \cdots \frac{1}{n+k}\right) \left(\frac{x}{2}\right)^{2k}$$

is a Bessel Function or Cylindrical Harmonic of the nth order and

second kind. γ is **Euler's Constant**. N (x) is sometimes referred to as
Neumann's or **Weber's Bessel Functions of the Second Kind.**

Bessel Inequality

$$|A| + |B| \geqslant |A + B|$$

where A and B are two vectors.

Bessel's Integral Equation

A **Bessel Function** of nth order can be represented by the integral
equation

$$J_n(x) = \frac{1}{\pi} \int_0^\pi \cos(nt - x \sin t)dt \qquad (n = 0, 1, 2 \dots)$$

Bessemer Process

A process used in the manufacture of steel whereby the impurities
present in pig iron are directly oxidized by a blast of air. The process
depends upon the oxidation of carbon to carbon monoxide and silicon
to silica which reacts with the metal to form a silicate slag. Phosphorus
impurities are more difficult to deal with and in the original Bessemer
process where the lining of the converter was ganister, a siliceous
material, a phosphorus bearing iron could not be used. However, if
the converter is lined with calcium oxide and magnesium oxide bound
together with tar and lime added at the end of the blow, the phosphorus
can be almost completely removed. This modification is the basis of the
Gilchrist Thomas process.

Betti's Reciprocal Theorem

If an elastic body is subjected to two systems of body and surface
forces, then the work that would be done by the first system of forces
in acting through the displacements due to the second system of
forces is equal to the work that would be done by the second system
of forces acting through the displacements due to the first system of
forces.

Betts' Process

An electrolytic process for refining lead. The electrolyte is a solution of lead silicofluoride and hydrofluosilicic acid. The lead is deposited on the cathode and the impurities remain on the anode.

Bhabba Scattering
Positron scattering by an electron.

Bial's Test
Take 2–3c.c. of urine and 4–5c.c. of Bial's Reagent which contains 1–1·5gm of orcin, 500c.c. of concentrated hydrochloric acid and 30 drops of a 1% solution of ferric chloride and heat until boiling commences. A green colour or the formation of a green precipitate indicates a pentose.

Bienaymé-Chebyshev Inequality
If $x_1, x_2 \ldots x_n$ are uncorrelated random variables which have a mean value $\bar{x}$ ($= \dfrac{1}{n} \displaystyle\sum_{i=1}^{n} x_i$) and which are from a distribution having a mean value m and standard deviation σ, then for any $\lambda > 0$,

$$P \left\{ (\bar{x} - m) \geqslant \frac{\lambda \sigma}{\sqrt{n}} \right\} \leqslant \frac{1}{\lambda^2}$$

where P represents probability.

Bingham Fluid
A substance which remains rigid under a shear stress until the magnitude of the stress exceeds the yield stress, whereupon the substance flows like a **Newtonian Liquid**.

Biot-Fourier Equation
An equation for the conduction of heat through a solid;

$$\frac{\partial Q}{\partial t} = \frac{\lambda}{C\rho} \nabla^2 T$$

where $\partial Q/\partial t$ is rate of flow of heat, λ the thermal conductivity and C the specific heat of the solid of density ρ. (*See also* **Fourier's Law**.)

Biot Number *See* Nusselt Number

Biot-Savart Relation *See* **Ampère's Law**

Biot's Law
The rotation of the plane of polarization of light propagated through an optically active medium is proportional to the length of the path, to the concentration (if the medium is a solution) and approximately inversely proportional to the square of the wavelength of the light.

Birch Reduction
A method of producing non-bezanoid cyclic ketones by reduction of a phenol with sodium in liquid ammonia in the presence of ethyl alcohol.

Birkeland-Eyde Method
The fixation of nitrogen from the air depends upon an endothermic reaction between oxygen and nitrogen and is therefore favoured by high temperatures. As the reaction is reversible, rapid cooling of the product, nitric oxide, is necessary in order to prevent dissociation. The method of Birkeland and Eyde, although of little practical importance today, was the first method used commercially and depended upon passing the air through an electric arc.

Bitter Figures
A method of viewing domains in a ferromagnetic material is to electrolytically polish the material to obtain a strain-free surface and then to spread a colloidal suspension of magnetite particles on the surface. The particles are drawn to regions of local maxima in the magnetization and a powder pattern or Bitter figure is obtained.

Bjerrum Theory of Molecular Spectra
In 1912 Bjerrum pointed out that the width of the infrared absorption bands of gases was of the order of magnitude to be expected on the assumption that they were due to molecular vibration in combination with molecular rotation. This doublet structure had already been observed in 1910 by A. Trowbridge and R. Wood but its significance had not been recognized.

Bjerrum's Treatment of Ion Association
In the calculation of the electric density in the vicinity of an ion in an electrolyte, Bjerrum suggested that certain difficulties associated with

the integration of a **Poisson equation** could be avoided by using the concept of the association of ions to form ion-pairs.

Black's Test
A test for β-hydroxy butyric acid in urine. The acid is first oxidized to diacetic acid which is then detected by ferric chloride.

Blagden's Law
The fact that a dissolved substance depresses the freezing point of water has been known for many years. In 1771 R. Watson observed that the time a solution took to freeze was proportional to the concentration. That the depression of freezing point is proportional to the concentration of the dissolved substance was observed by C. Blagden 1788 and F. Rudorf 1861.

Blanc Rule
A generalization that states: whereas succinic and glutaric acids yield cyclic anhydrides on pyrolysis, adipic and pimelic acids yield cyclic ketones. It has been used to determine whether oxygenated or unsaturated rings of compounds of unknown composition are five or six membered. Thus, when a five-membered ring is oxidized it gives glutaric acid which pyrolyses to an anhydride, whereas a six-membered ring gives adipic acid and hence a ketone. The rule should be used with caution as certain cases are known where, for example, substituted adipic acid gives rise to a seven-membered cyclic anhydride.

Bloch's Functions
The solutions of the Schrödinger equation with a periodic potential are of the form

$$\psi = u_k(r)\, e^{jk.r}$$

where u is a function depending in general on k, which is periodic in x, y, z with the periodicity of the potential; that is, with the period of the lattice. This result was known earlier as **Floquet's Theorem**.

Bloch's $T^{3/2}$ Law for Magnetization
At very low temperatures ($T/T_C \ll 1$, where T_C the **Curie Temperature**), the spontaneous magnetization M_S may be expressed as a function of temperature by

$$M_{S,\,T} = M_{S,\,0}\,(1 - CT^{3/2})$$

where C is a constant equal to

$$\frac{0 \cdot 0587}{a\,S}\left(\frac{k}{2SJ}\right)^{3/2}$$

Here, $a = 1$, 2 and 4 respectively for simple cubic, body-centred cubic and face-centred cubic structures, whose ions have spin S. J is an exchange integral relating the energy of interaction with the magnitude and direction of the spins of neighbouring ions.

This law is in quite good agreement with observation in the very low temperature region but at somewhat higher temperatures a T^2 term replaces the $T^{3/2}$ term.

Bloch's Theorem of Superconductivity
Bloch proved that the lowest state of a quantum mechanical system in the absence of a magnetic field can carry no current.

Bloch Wall
This term denotes the transition layer which separates domains magnetized in different directions.

Blondel-Rey Law
A light source, flashing at low frequency (less than 5 hertz) and having brilliance B_0 during the flash, has an apparent brilliance B given by

$$B = \frac{B_0 t}{a + t}$$

where t is the time of duration of the flash and a is a constant (equal to approximately $0 \cdot 2$ sec when B is near the threshold for white light). See also **Talbot's Law.**

Boas' Test
For the detection of free hydrochloric acid in gastric juices, an alcoholic solution of resorcinol and sucrose gives a red colour.

Bode's Law
Bode stated that, if the distance of Mercury from the Sun is taken as 4, the corresponding distances of the other planets would be given, in the

same units, as: Venus 4 + 3, the Earth 4 + 6, Mars 4 + 12, Asteroids 4 + 24, Jupiter 4 + 48, Saturn 4 + 96 and Uranus 4 + 192. Neptune, however, which should lie, according to Bode's law at a distance of 388 from the Sun was found to be in fact, 300 units away.

Bodroux-Chichibabin Reaction

A method of synthesizing aldehydes in which reaction occurs between ethyl orthoformate and a Grignard reagent (*see* **Grignard Reaction**) containing the alkyl group required in the aldehyde. The acetal formed is hydrolysed in acid solution to give the aldehyde.

$$HC(OC_2H_5)_3 + RMgX \rightarrow RCH(OC_2H_5)_2 + C_2H_5OMgX$$
$$H^+ \downarrow H_2O$$
$$2C_2H_5OH + RCHO$$

Boettger's Test

A test for the presence of saccharoses based on the reduction of bismuth nitrate to metallic bismuth in alkaline solution.

Bohr-Breit-Wigner Theory

A nuclear reaction occurs in two stages. In the first, the bombarding particle is absorbed by the target nucleus to form a high energy unstable compound nucleus. In the second, this nucleus emits radiation to leave a residual lower energy nucleus known as a recoil nucleus.

Bohr Magneton *See* Appendix

Bohr Radius *See* Appendix

Bohr's Correspondence Principle

The behaviour of the electrons in an atomic system must approach more and more that predicted by classical physics the higher the quantum number of the orbit.

Bohr-Sommerfeld Atom

Rutherford, in 1911, as a consequence of experiments relating to the scattering of α-particles by a metal foil postulated that the atom was composed of a very small positively charged nucleus surrounded at a relatively large distance by a number of negatively charged electrons. To account for the fact that the electrons do not fall into the nucleus as a result of electrostatic attraction, Rutherford found the attractive force inward was balanced by the outward centrifugal force. It was pointed out, however, that such a system cannot be stable as a particle moving at constant speed in a circle has an acceleration towards the centre. If the particle is an electron it should radiate energy because the electromagnetic field associated with an accelerating electron is continually changing. N. Bohr suggested that there were certain circular orbits in which an electron could move without emitting or absorbing energy. (*See* **Bohr's Theory of Spectral Emission**.) The number of orbits or energy levels is determined by the quantum condition that the angular momentum of the electron is an integral multiple of $h/2\pi$. This theory was modified by A. Sommerfeld, who pointed out that the periodic motion of a particle subjected to a central force is elliptical with the central body at the focus of the ellipse. Thus electron orbits should be elliptical, with the nucleus of the atom being situated at a focus of the ellipse. When the electron is at its nearest point to the nucleus, its acceleration is greater than when it is at its farthest point from the nucleus, and hence, according to the theory of relativity, the mass of the electron changes. The electron does not, therefore, continue in a fixed plane but precesses and the ellipse is not complete. This view of the atom is completely pictorial for it tacitly assumes that both the momentum and the position of an electron are known, a state of affairs at variance with the **Heisenberg Uncertainty Principle**. The major postulate of quantization in $h/2\pi$ units, however, has been vindicated by the methods of wave mechanics.

Bohr's Theory of Spectral Emission

Bohr in 1913, by applying the principles of quantum mechanics to atomic structure problems, was able to suggest a theory of the origin of spectral lines. Thus, for the hydrogen or positively charged helium atom, he was able to restrict the emitted radiation to one of the frequencies specified by

$$\nu = R\left(\frac{1}{T_2^{\,2}} - \frac{1}{T_1^{\,2}}\right)$$

where ν is the frequency, T_2 and T_1 are positive integers and

$$R = \frac{2\pi^2\, e^4\, Z^2\, \mu}{h^3}$$

31

where e is the electronic and Z the nuclear charge, μ is the reduced mass and h is Planck's Constant (*see* **Ritz Combination Principle**). His two major conclusions were stated in his original paper as:

(1) An atomic system can, and can only, exist permanently in a certain series of states corresponding to a discontinuous series of values for its energy, and, consequently, any change in the energy of the system, including emission and absorption of electro-magnetic radiation, must take place by a complete transition between two such states, denoted as the 'stationary states' of the system.

(2) The radiation absorbed or emitted during a transition between two stationary states is monochromatic and of frequency given by

$$h\nu = E_n - E_m$$

where E_n and E_m are the energies of the two different states.

Bohr's Theory (of Light Absorption)

Bohr regarded the absorption of light as the reverse of emission. He attributed it to the transition of an electron from a lower to a higher orbit. The energy necessary for this transition is furnished by a light quantum of exactly the correct magnitude.

Bohr's Theory (of Resonance Radiation)

Resonance radiation is produced by the return of an electron to the same ground state as that from which it had been excited. Thus the Hg atom is capable in its normal state of absorbing a quantum of radiation corresponding to wavelength 2537 Å by which the electron is carried from the inner 1^1S_0 orbit to the 2^3P_1 orbit. The return of an electron from this orbit to the inner one gives rise to resonance radiation of the same wavelength as the light absorbed.

Bohr-Van-Leeuwen Theorem

These workers have shown that in any dynamical system to which classical statistical mechanics can be applied, the paramagnetic and diamagnetic susceptibilities cancel, so that, classically, for thermal equilibrium, one always has zero susceptibility.

Bohr-Wheeler Theory of Fission

A theory of fission based on a liquid-drop model of the nucleus in which there are Coulombic and surface tension forces. (*See also* **Weizsäcker's Formula.**)

Boltzmann Distribution Law *See* **Maxwell-Boltzmann Distribution Law**

Boltzmann Equation
The entropy of a system of particles is proportional to the logarithm of the probability of its macroscopic state. The constant of proportionality is known as the **Boltzmann Constant**. (*See* **Appendix**.)

Boltzmann H Theorem
The **Boltzmann H function** is defined as

$$H = \iint f \ln f \, d\mathbf{v} \, d\mathbf{r}$$

where integration is over all position and momentum space. f is a distribution function and $\mathbf{v}$ and $\mathbf{r}$ are velocity and position vectors respectively. Then, Boltzmann's H theorem states that H is a decreasing function with time unless f is Maxwellian in which case H remains constant. This is analogous to the thermodynamic statement that the entropy of a system tends to a maximum, and, except for an additive constant, the H function equals the negative of entropy divided by Boltzmann's constant.

Boltzmann Law of Radiation (Stefan Boltzmann Law)
Stefan induced an experimental law for the intensity of the total radiation from a black-body at various temperatures. This law was subsequently derived theoretically by Boltzmann. The law states that the complete emission W of a black body is proportional to the fourth power of the absolute temperature.

$$W = aT^4$$

where a is a constant.

Boltzmann's Ratio
Where the ion of a compound can have two energy levels not widely separated, the distribution of ions (n_1, n_2) over these levels is controlled by Boltzmann's ratio

$$\frac{n_1}{n_1} = \exp\left(\frac{\Delta E}{RT}\right)$$

where ΔE is the energy difference per mole.

33

Boltzmann Statistics *See* **Maxwell-Boltzmann Statistics**

Boltzmann's Transport Equation
An equation expressing the variation $(f\text{-}f_0)$ of a distribution function
for charged carriers from its equilibrium value f_0 in the presence of a
thermal gradient and fields such as electric and magnetic fields. If $\mathbf{v}$ is
the velocity of the carriers, $\mathbf{p}$ their momentum and $d\mathbf{p}/dt$ the force on
them at time t, and τ ($\mathbf{p}$) is a relaxation time for the collision processes,
then for elastic scattering of the carriers,

$$\frac{\partial f}{\partial t} + \mathbf{v}\,\nabla f + \frac{d\mathbf{p}}{dt}\,\nabla_p f = -\frac{f-f_0}{\tau\,(\mathbf{p})}$$

For steady-state processes $\partial f/\partial t = 0$.

Bolzano-Weierstrass Theorem
A bounded sequence of numbers always contains a convergent sequence.
(A sequence is bounded if there is a number A such that for all values
of n, $|x_n| \leqslant A$). For instance, consider the bounded but non-convergent
sequence $\{x_n\}$ where $x_n = (-1)^n$. This has a sub-sequence $\{x_{2n}\}$ which
converges to 1.

Bolza, Problem of
The general problem of finding the class of curves $y_i = y_i(x)$ such that

$$\int_{a(z)}^{b(z)} F_i(x, y, y')\,dx + G_i(z)$$

are minima subject to constraints of the form

$$\Phi_j(x, y, y') = 0.$$

For fixed points, $G_i(z) = 0$, and the problem becomes an isoperimetric
one; for example, the problem of finding the path between two points
of a particle of pre-assigned initial velocity under constant acceleration
due to gravity.

Bonnet's Form
If $f(x)$ and $\phi(x)$ are integrable between a and b and if $\phi(x)$ is a positive
decreasing function of x, then a number ξ exists such that $a \leqslant \xi \leqslant b$

34

and

$$\int_a^b f(x)\,\phi(x)\,\mathrm{d}x = \phi(a)\int_a^\xi \mathrm{f}(x)\,\mathrm{d}x.$$

Boolean Algebra
An algebra of sets used to enunciate logical propositions and their consequences. The algebra has two binary operations called addition and multiplication. The algebra may be used to represent binary logic and can be applied to computer logic, switching networks and so on. *See also* **De Morgan's Rules** *and* **Venn Diagram**.

Boole's Theorem
The linear differential equation

$$\sum_{k=0}^{n} f_k(\vartheta)\,\mathrm{e}^{kz}\,.\,u = 0$$

where $x = \mathrm{e}^z$ and $xD = \vartheta$ where $D = \mathrm{d}/\mathrm{d}x$ has as many independent solutions in ascending power series as there are simple linear factors in $f_0(\vartheta)$ and as many descending developments as there are factors of a like nature in $f_n(\vartheta)$.

Bordini Effect
When metals possessing a close-packed crystal structure are subjected to an oscillating stress, there is a peak in the internal friction at a particular frequency of applied stress. It is considered to be due to dislocations, which run parallel to close-packed rows, oscillating from one equilibrium position to the adjacent equilibrium position.

Born Approximation
A method of obtaining the asympotic form for a scattered wave (e.g. in quantum mechanics) depending on performing a sequence of iterative integrations.

Born Equation
The heat of solvation of an ion is given by

$$\Delta H = -\frac{Nz^2e^2}{2r}\left(1 - \frac{1}{\epsilon_r}\right)$$

35

where N is the Avogadro number, z is the valency of the ion, r is the effective radius of the ion and ϵ_r is the dielectric constant of the solvent.

Born-Haber Cycle

An imaginary cycle for a crystalline halide used to calculate its lattice energy. The steps in the cycle are:

(1) Heat is removed from the crystal to reduce its temperature to absolute zero.
(2) The crystal is broken into free stationary ions.
(3) Electrons are transferred from the negative ions to positive ions to produce neutral atoms.
(4) The halogen atoms are allowed to recombine to form diatomic atoms and
(5) the gas is brought to room temperature and pressure.
(6) The metal vapour is also returned to room temperature and
(7) condensed to the metal.
(8) The metal and the halogen are allowed to combine chemically, thus completing the cycle.

If the energy associated with each step except (2) is known, the lattice energy can be calculated.

Borrmann Effect (or Campbell-Borrmann Effect)

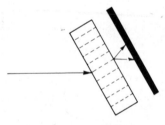

Borrmann Effect

The anomalous transmission of x-rays when a single crystal of high perfection is placed in a reflecting position in a monochromatic x-ray beam. The transmitted and diffracted beams emerge displaced and with anomalously high intensity. Adsorption is reduced because standing waves are established in the crystal with nodes at the atomic sites the displacement is because energy flow in the crystals is along the reflecting planes. The effect is the basis of an x-ray topography method of photographing dislocations in solids.

Bosanquet's Law

The resistance of an electric circuit is defined as the ratio of the electromotive force to the current (**Ohm's Law**). In a magnetic circuit the ratio of the magnetomotive force F_m to the magnetic flux Φ is known as the reluctance R. Bosanquet's Law states that

$$\Phi = \frac{F_m}{R} \text{ in analogy to Ohm's Law.}$$

Bose-Einstein Condensation

As the temperature of an assembly of Bose-Einstein particles tends to zero, the particles are able to fall into the zero energy state. This is called Bose-Einstein Condensation and is condensation into momentum space rather than condensation into coordinate space which occurs for liquefaction from the gas phase. The behaviour is in contrast to that of an assembly of Fermi-Dirac particles where only one particle per state including the zero energy state is allowed because of the **Pauli Exclusion Principle**, and where, at zero temperature, particles fill the states up to the **Fermi energy level**.

Bose-Einstein Statistics

The study of the probability of occupation of energy states by indistinguishable non-interacting independent particles in a quantized system. These particles are allowed to be arranged among the energy states without limit to the number of particles per state, and this fact differentiates Bose-Einstein statistics from **Fermi-Dirac Statistics**. The probability that an energy state E is occupied is given by the **Bose-Einstein distribution law**;

$$f(E) = \left\{ e^{(E - \mu)/kT} - 1 \right\}^{-1}$$

where μ is a constant. Particles which are subject to Bose-Einstein statistics are called Bose-Einstein particles or **bosons**.

Bouguer's Law *See* Lambert's Law

Bouveault-Blanc Reduction

When aldehydes, ketones or esters are reduced with sodium and ethanol, monohydric alcohols are formed, e.g.

$$RCHO + 2H \rightarrow RCH_2OH$$

37

Bower-Barff Process
A method of reducing the corrosion of iron or steel wherein the metal is raised to a red heat and is then subjected to superheated steam.

Bow's Notation
A method of notation of forces acting at a point, wherein the spaces between forces are lettered or numbered, any force being defined by the letters or numbers on each side of it.

Boyle's Law (Marriotte's Law)
At a constant temperature the volume of a definite mass of gas is inversely proportional to the pressure, i.e.

$$pV = \text{constant.}$$

Real gases deviate appreciably in behaviour from this equation as might be expected as the law may be deduced from kinetic theory which considers the molecules of the gas as point masses having no attraction for one another.

Boyle's Temperature (Point)
In general, real gases do not obey **Boyle's Law**. The relation between the pressure and the volume of a real gas may be expressed by a power series

$$pV = A + Bp + Cp^2 \ldots$$

The coefficients A, B, C, etc., are called the virial coefficients. The first virial coefficient A is equal to RT where T is the absolute temperature and the effect of the third and higher virial coefficients is small except at high pressures. The deviation of the behaviour of the gas from Boyle's Law is therefore determined by the second virial coefficient B which is temperature dependent (the values of B at various temperatures for different gases are given in the figure). The temperature at which B becomes zero is known as the Boyle Temperature, for, at this temperature, providing the higher virial coefficients are negligible, the gas will obey Boyle's Law up to comparatively high pressures.

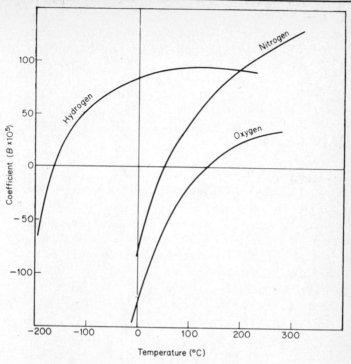

Boyle's Temperature

Brackett Series *See* **Balmer Series**

Bragg-Kleenan Rule
The atomic stopping power of a material to charged particles is proportional to the square root of the material's atomic weight. Atomic stopping power is additive and thus can be obtained for complex molecules.

Bragg-Pierce Law
The atomic absorption coefficient of an element for x-rays is given by

$$\mu_a = C\,Z^4\,\lambda^3 \text{ (sometimes quoted } \mu_a = C'Z^5\,\lambda^{7/2})$$

where Z is the atomic number, λ is the wavelength of the x-rays and the constant C (or C') varies with λ.

39

Bragg's Law

Let the distance between lattice planes in a crystal be d, let the wavelength of a homogeneous beam of x-rays be λ and let the glancing angle be θ. Then for diffraction.

$$n\lambda = 2d \sin \theta \ (n \text{ an integer})$$

This relation, known as the **Bragg equation**, enables the crystal parameters to be determined.

Bravais Lattice

A crystal is built up by the repetition by translation in three dimensions of a structural unit. The nature of the basic pattern is apparent if each of these structural units is replaced by a point. In this infinite three-dimensional array each point is in the same environment as any other point. Bravais showed that there can only be fourteen of these space lattices. The unique lattices are shown below.

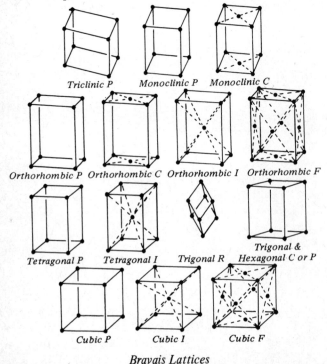

Triclinic P Monoclinic P Monoclinic C

Orthorhombic P Orthorhombic C Orthorhombic I Orthorhombic F

Tetragonal P Tetragonal I Trigonal R Trigonal & Hexagonal C or P

Cubic P Cubic I Cubic F

Bravais Lattices

40

Bravais' Law

Those crystal forms which tend to occur most frequently are those with faces parallel to planes of smallest reticular area. (Reticular area is inversely proportional to the interplanar spacing.) *See also the* **Donnay-Harker Principle.**

Bredt Rule

Reactions of the **Diels-Alder** type with cyclopentadiene give cyclanes having a methylene bridge across a cyclohexane ring. The corresponding alcohol does not hydrate as double bonds cannot exist in the bridge.

Breit-Wigner Formula

If a nucleus is excited to a well-defined single excitation level by an incoming particle n, for instance a neutron, with the emission of a particle p, then the reaction cross-section is given by

$$\frac{\pi g}{k^2} \frac{\Gamma_n}{(E - E_r)^2 + (\Gamma/2)^2} \Gamma_p$$

where

$$g = \frac{2J + 1}{(2I + 1)(2s + 1)}.$$

J is the spin of the compound nucleus, I the spin of the target nucleus and s the spin of the incoming neutron n. Γ_n and Γ_p are the neutron and particle energy widths and Γ is the total energy width of the compound nucleus (proportional to the disintegration constant). E_r is the energy at the peak of the resonance, E the total energy (of the target and incoming neutron) and k the wave vector associated with the incident neutron.

Brewster *See* Appendix

Brewster's Fringes

Interference fringes observed when light from an extended source passes through two etalons (pairs of parallel semi-silvered glass plates with an air-gap between) which are inclined at a small angle. There is a path difference established between light waves reflected in the two etalons, which must be of equal thickness or the ratio of whose thicknesses is an integer. Usually, transmission fringes are observed, but the reflection fringes in a Jamin interferometer are a particular case of Brewster's fringes.

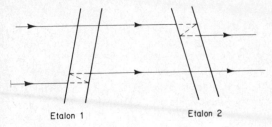

Etalon 1   Etalon 2

Brewster's Fringes

Brewster's Law

When a beam of light is reflected from a surface, polarization may occur. As the angle of incidence is increased from normal to grazing the polarization increases, passes through a maximum and then decreases. The angle of maximum polarization is known as the angle of polarization or **Brewster angle** and varies with the nature of the substance. Brewster discovered that the refractive index was equal to the tangent of the angle of polarization. This law leads to the result that the sum of the angle of incidence and angle of refraction at maximum polarization is equal to a right angle.

Brianchon's Theorem

For any hexagon whose sides are tangent to a conic, the diagonals connecting opposite vertices intersect in a point (or are parallel).

Bridgman Effect

The absorption or liberation of heat when an electric current passes through an anisotropic crystal and arises from non-uniformity in the current distribution. It is additional to the **Peltier** and **Thomson Effects.**

Bridgman Relation

In a metal or semiconductor

$$PK = QT$$

where P is the Ettingshausen coefficient, K is the total thermal conductivity and Q is the Nernst-Ettingshausen coefficient. (*See* **Ettingshausen** and **Nernst Effects.**)

Bridgman-Stockbarger Method

A method of single crystal growth in which a melt, contained in a capsule, is lowered through a temperature gradient about the freezing point. The method either involves a two-zone furnace, in which case the upper zone is at a temperature above, and the lower zone at a temperature below, the freezing point, or a single zone furnace, when use is made of the drop in temperature towards the outside of the furnace. By either tapering the capsule or constricting it at the lower end, a single crystal is selected. A horizontal modification of the method can also be used.

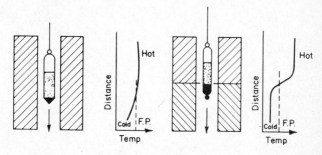

Bridgman-Stockbarger Method

Briggsian Logarithm

If $y = 10^x$, x is termed the Briggsian Logarithm of y. *See also* **Brig (Appendix)**.

Brillouin Function

In a paramagnetic substance the magnetization can be shown to be

$$M = NgJ\mu_B \, B_J(x)$$

where N is the number of atoms per unit volume each with a total angular momentum quantum number, J, and a spectroscopic splitting factor g (**Lande Splitting Factor**) and μ_B is the **Bohr Magneton**, and

$$x = \frac{gJ\mu_B \, H}{kT} .$$

B_J is the Brillouin function

$$B_J = \frac{2J+1}{2J} \coth \frac{(2J+1)x}{2J} - \frac{1}{2J} \coth \frac{x}{2J} .$$

43

Brillouin Scattering

The scattering of electromagnetic waves at optical frequencies by sound waves (acoustic phonons) at microwave frequencies and analogous to **Raman Scattering**.

Brillouin Zone

If the reciprocal lattice of a crystal is obtained, but with periodicity of the lattice of $2\pi/d_{hkl}$ instead of $1/d_{hkl}$, and the space divided into identical cells in the same way as the real lattice is divided to obtain the **Wigner-Seitz Cell**, then the cells so obtained are called Brillouin zones. These zones are very important in the theory of energy bands in crystals because, by **Bragg's Law**, the zone may be said to define a momentum or **k** space in which **k** is the wave vector for the electrons. If at the reciprocal lattice points there is strong reflection of electrons or x-rays (this will be so unless the so-called structure factor used in x-ray analysis is zero), then there will be an energy gap at each face of the Brillouin zone. (*See also* **Ewald Sphere** *and* **Jones Zone**.)

Brinell's Hardness Test

The hardness of a material is measured by the area of indentation produced by a standard steel ball of 1 cm diameter under prescribed conditions of loading (usually 3000 kg static pressure for steel and 500 kg for softer metals). The **Brinell number** is given by

$$\frac{p}{\pi t D} = \frac{p}{\pi D \{D/2 - \sqrt{(D^2/4 - d^2/4)}\}}$$

where p = pressure in kg,
$\quad t$ = depth of indentation in mm,
$\quad D$ = diameter of ball in mm,
$\quad d$ = diameter of indentation in mm.

Brin's Process

Formerly oxygen was obtained industrially by this process from barium oxide which was first oxidized in air to barium dioxide and then heated, when oxygen was evolved and the oxide regenerated. This method has now been supplanted by liquid-air processes.

Bromwich's Expansion Theorem (*see also* **Heaviside's Expansion Theorem**)

If
$$u = \frac{F(p)}{f(p)} \, (Gt)$$

where G is a constant, the solution is

$$u = G \left\{ N_0 t + N_1 + \sum \frac{F(a)}{a^2 \, f'(a)} \, e^{at} \right\}$$

where N_0 and N_1 are defined by

$$\frac{F(p)}{f(p)} = N_0 + N_1 p + N_2 p^2 + \ldots$$

a is any root of $f(p) = 0$, $f'(p)$ is the first derivative of $f(p)$ with respect to p and the summation is over all the roots a. The solution reduces to $u = 0$ when $t = 0$.

Brønsted-Lowry Definition of acids and bases
According to these workers, an acid is defined as a substance with a tendency to lose a proton and a base as a substance with a tendency to gain a proton.

Such a property of any substance cannot manifest itself unless the solvent molecules are themselves able to act as proton acceptors or donors.

Brownian Movement
In 1827 the botanist, R. Brown, observed under the microscope that pollen grains in water were in a continual state of haphazard motion. The erratic motion has been observed for all kinds of small particles suspended in various media. The phenomenon is explained by the continual buffeting of the suspended particles by the molecules of the medium in which they are suspended.

Brunauer, Emmett and Teller (B.E.T.) Adsorption Equation
Whereas Langmuir's equation (*see* **Langmuir Adsorption Isotherm**) assumes an adsorbed layer which is monomolecular, the B.E.T. equation assumes multimolecular adsorption with each separate layer obeying

the Langmuir equation. The adsorption equation becomes

$$\frac{V}{V_m} = \frac{c\,p/p_s}{(1 - p/p_s)\,(1 - p/p_s + cp/p_s)}$$

where V is the volume of gas adsorbed, V_m is the volume of gas required to cover the entire surface with a monomolecular layer, p is the pressure of the gas and p_s the saturation pressure. c contains the nett heat of adsorption:

$$c = c_1 \exp\left(\frac{Q_A - Q_L}{RT}\right)$$

where Q_A is the average heat of adsorption, Q_L the heat of liquefaction of the vapour and c_1 a constant of the system.

Bucherer Reaction

A naphthol may be reversibly converted into a naphthylamine by the action of ammonia in the presence of an aqueous solution of a sulphite.

Buckingham's Π Theorem

Let $a, \beta, \gamma, \ldots$ be n measurable and dimensional quantities expressible in terms of m primary quantities.

Buckingham's theorem states that if

$$f(a, \beta, \gamma, \ldots) = 0$$

is to be a complete equation, the solution has the form

$$F(\pi_1, \pi_2, \ldots) = 0$$

where the π's are the $(n - m)$ independent products of the arguments $a, \beta, \gamma, \ldots$ which are dimensionless in the fundamental units.

46

Budan's Theorem

For a real algebraic equation of the form

$$f(x) = a_0 x^n + a_1 x^{n-1} + \ldots a_n = 0$$

where $a_0 \neq 0$

if $N(x)$ is the number of changes of sign in the sequence of derivatives $f(x), f'(x), f''(x) \ldots f^n(x)$ where vanishing derivatives are disregarded, then the number of real roots located between two real numbers a and $b > a$, where a and b are not roots themselves, is either $N(a) - N(b)$ or is less than $N(a) - N(b)$ by a positive even integer.

Budde Effect

The expansion observed on exposing halogen vapours to light caused by the rise in temperature resulting from heat evolved in the recombination of atoms.

Burger-Dorgelo-Orstein Rule

For small multiplet splitting in atomic spectra when **Russell-Saunders Coupling** is applicable, the sum of the intensities of all the lines of a multiplet which belong to the same initial or final state is proportional to the statistical weight of $2J + 1$ of the initial or final state respectively, where J is the quantum number for the total angular momentum of the electrons.

Burgers' Circuit

A method of determining whether an imperfection is contained in a

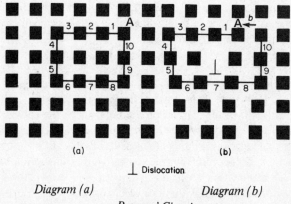

⊥ Dislocation

Diagram (a)   *Diagram (b)*

Burgers' Circuit

47

crystal lattice. Consider the perfect two-dimensional structure in the figure (a). The circuit marked with the arrows consists of ten moves to go from A back to A; e.g. $1 \rightarrow 2 \rightarrow 3$ and so on. In the figure (b), a dislocation is contained in the area of the Burgers' Circuit and it now takes more than ten moves to go from A back to A.

Burgers' Vector
The vector necessary to close **Burgers' Circuit**. It is at right angles to the dislocation for a pure edge dislocation and parallel to the dislocation for a screw dislocation.

Burnside's Formula
A formula for double integration:

$$\int_{-1}^{+1} \int_{-1}^{+1} f(x, y) \, dx \, dy = \tfrac{40}{49} \{ f(a, 0) + f(0, a) + f(-a, 0) + f(0, -a) \}$$

$$+ \tfrac{9}{49} \{ f(b, b) + f(b, -b) + f(-b, b) + f(-b, -b) \}$$

where $a = \sqrt{\tfrac{7}{15}}$, $b = \sqrt{\tfrac{7}{9}}$ and $f(x, y)$ is a polynomial of degree 5 at most.

Bury's Theory
A theory, based on chemical considerations, of the electronic structure of atoms. It included the supposition that a new outer sheath of electrons can commence to form while the inner ones are still incomplete.

Burstein Effect
The variation in width of the optical energy gap of certain semiconductors with the amount of doping. Usually, the optical gap is between the highest energy state in the valence band and the lowest state in the conduction band; i.e. the smallest energy difference required for excitation of electrons from the valence to the conduction band. In some semiconductors having a very large carrier density, the first unoccupied states lie well within one of the bands and this position is a function of doping.

Buys Ballot's Law

In meteorology the law giving the direction of rotation of cyclones and anticyclones. In the northern hemisphere with wind arriving from the rear of the observer the pressure is lower on the left-hand side rather then on the right; the converse is true in the southern hemisphere.

C

Cailletet and Mathias Law
The mean of the density of a liquid and its saturated vapour at the same temperature is a linear function of temperature. The densities of the liquid and of saturated vapour in equilibrium with it are known as orthobaric densities and if ρ_t is the arithmetical mean at any temperature t then

$$\rho_t = \rho_0 + at$$

where ρ_0 and a are constant for a given substance.

Callendar's Equation
An equation representing the behaviour of superheated steam below the critical point and away from the liquid-vapour region. It is:

$$V - b = \frac{RT}{p} - \frac{a}{T^n}$$

where $n = \frac{10}{3}$ and a and b are constants.

Calzecchi-Onesti Effect
The conductivity of loosely aggregated metallic powders changes under the action of applied electromotive forces.

Campbell's Formula
A formula giving the propagation constant of a loaded transmission line in terms of the constants of the unloaded line.

$$\cosh \gamma = \cosh Pd + \frac{Z_1}{2Z_0} \sinh Pd$$

where γ = propagation constant of the loaded line per d miles,
P = propagation constant of the unloaded line per mile,
Z_0 = characteristic impedance of unloaded line,
Z_1 = impedance of each loading coil.

Cannizzaro Reaction

This reaction is an example of disproportionation. When an aldehyde having no a hydrogen atoms is heated with concentrated aqueous caustic soda an alcohol is formed

$$2HCHO + NaOH \rightarrow H.COONa + CH_3OH$$

All aldehydes may be made to undergo this reaction in the presence of aluminium ethoxide.

Carathéodory's Principle

In the neighbourhood of any arbitrary point P_0, there are points which are not accessible from P_0 along curves which are solutions of the **Pfaffian Differential Equation**, if, and only if, the equation is integrable, For an adiabatic reversible thermodynamic cycle, one may write

$$đQ = A\,dx + B\,dy + C\,dz$$

where x, y and z are thermodynamic variables and Q is heat. From the second law of thermodynamics, there exist, neighbouring the state P_0 of a physical system, other states which are not accessible along adiabatic paths from P_0. Therefore, the equation must be integrable and the introduction of an integrating factor indicates the existence of an entropy function.

Carnot's Cycle

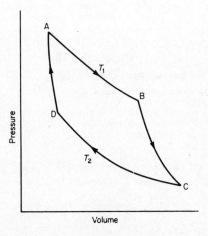

Carnot's Cycle

51

A sequence of operations forming the working cycle of an ideal heat engine of maximum thermal efficiency. It consists of isothermal expansion AB, adiabatic expansion BC, isothermal compression CD and adiabatic compression DA as shown in the diagram. Application of the First Law of Thermodynamics leads to an equation for the efficiency of any reversible machine:

$$\frac{Q_1 - Q_2}{Q_1} = \frac{T_1 - T_2}{T_1}$$

where Q_1 is the heat absorbed at the temperature T_1 and Q_2 is that given up again at the lower temperature T_2.

Carnot's Theorem
All machines working reversibly between the same temperature of source and sink have the same efficiency. That is to say, provided the machine functions reversibly, its efficiency is independent of the nature of the working fluid employed or the mode of operation.

Cassini, Ovals of
Curves giving the locus of points having the product of their distances from $(-a, 0)$ and $(a, 0)$ as c^2 and having the form

$$(x^2 + y^2 + a^2)^2 - 4a^2x^2 = c^4$$

Castigliano's Theorems
I. If U is the strain energy (*see* **Clapeyron's Theorem**) when a body is deformed, s_i are the displacement components and F_i are the corresponding force components, then

$$s_i = \frac{\partial U}{\partial F_i}$$

II. If the prescribed values of the displacement components are denoted s_i' then the unprescribed forces F_i are such that $(U - \Sigma\, s_i' F_i)$ is a minimum.

Castner-Kellner Process (for the Manufacture of Sodium Hydroxide)
A chemical process of pedagogical importance involving the use of a

mercury cathode in which the sodium liberated from the electrolysis of brine forms an amalgam from which the caustic soda is produced by the action of water.

Castner Process (for the Preparation of Sodium)
A process for the manufacture of sodium by the electrolysis of molten caustic soda. Due to the anodic formation of water which reacts with the sodium produced at the cathode the method is inefficient and has been replaced by other methods depending upon the electrolysis of brine (*see* Downs Process).

Cauchy-Riemann Equation
A function $f(x) = u(x, y) + jv(x, y)$ is analytic if the Cauchy Riemann equations

$$\frac{\partial u}{\partial x} = \frac{\partial v}{\partial y}$$

$$\frac{\partial u}{\partial y} = -\frac{\partial v}{\partial x}$$

hold.

Cauchy's Boundary Conditions
Boundary conditions involving the relation between the value and the normal gradient of the required solution on the boundary.

Cauchy-Schwarz Inequality

$$\left| \sum_j a_j b_j \right|^2 \leqslant \left(\sum_j |a_j|^2 \right) \left(\sum_j |b_j|^2 \right)$$

Cauchy's Condensation Test
If $f(n)$ is a positive decreasing function of n, and a is any positive integer greater than 1, the two series $\Sigma f(n)$ and $\Sigma a^n f(a^n)$ are both convergent or both divergent.

Cauchy's Dispersion Formula

$$n = A + \frac{B}{\lambda^2} + \frac{C}{\lambda^4}$$

where n is the refractive index and A, B and C are constants. As A, B and C are determined from observations of n at widely different wavelengths the formula represents the dispersion of most substances with considerable accuracy, although the theory used by Cauchy, the elastic solid theory of light, is not true.

Cauchy's Frequency Distribution

A frequency distribution for variable x given by

$$f(x) = \frac{1}{\pi a} \; \frac{1}{1 + [(x - \lambda)/a]^2}$$

where the distribution is symmetric about λ which is the median but not the mean. a is a measure of the spread of the distribution.

Cauchy's Integral Formula

If $f(z)$ is analytic inside and on a closed contour C, and if a is a point, then

$$\oint \frac{f(z)}{z - a} \, dz = 2\pi j \, f(a).k$$

where $k = 1$ if a is inside C, $k = 1/2$ if a is on C and $k = 0$ if a is outside C.

Cauchy's Integral Theorem

If $f(z)$ is analytic in the closed region bounded by a simple closed contour then $\oint f(z) \, dz = 0$.

Cauchy's Mean

Arithmetic mean; If $a_n \rightarrow a$ as $n \rightarrow \infty$, then $(a_1 + a_2 + \ldots a_n)/n \rightarrow a$
Geometric mean; If $a_n \rightarrow a > 0$ as $n \rightarrow \infty$ and $a_n > 0$, then $(a_1 a_2 \ldots a_n)^{1/n} \rightarrow a$.

Cauchy's Mean Value Theorem

If $f(x)$ and $\phi(x)$ are continuous in the range $a \leqslant x \leqslant b$, and $f'(x)$ and

$\phi'(x)$ exist and do not vanish simultaneously in the range $a < x < b$, then there exists a point c such that $a < c < b$ where

$$\frac{f(b) - f(a)}{\phi(b) - \phi(a)} = \frac{f'(c)}{\phi'(c)}$$

Cauchy's Rule for Series

$$\sum_{n=0}^{\infty} \left(\sum_{j=0}^{n} a_j b_{n-j} \right) = \left(\sum_{j=0}^{\infty} a_j \right) \left(\sum_{j=0}^{\infty} b_j \right)$$

provided the three series are convergent.

Cauchy's Tests for Convergence
(1) In

$$\sum_{n=1}^{\infty} u_n = u_1 + u_2 + \dots$$

   if

$$\lim_{n \to \infty} |u_n|^{1/n} < 1$$

u_n is absolutely convergent.

(2) If $f(x)$ be a steadily decreasing positive function such that $f(n) \geqslant a_n$, then

$$\sum a_n \text{ is convergent if } \int_m^{\infty} f(x) \, dx \text{ is convergent.}$$

Cauchy's Theorem (for a Square Matrix)
The adjugate of a square matrix is that matrix whose (i, j)th element is the cofactor of the (j, i)th element in $|A|$, the determinant of A; the cofactor is $(-1)^{i+j} \times$ the minor, and the minor is formed from the

square matrix A by suppressing the ith row and j^{th} column. Then, Cauchy's theorem states that if A is a square matrix of order n,

$$| \text{adj } A | = | A |^{n-1}$$

Cauchy's Theorem (for the Existence of a Limit)

The necessary and sufficient condition for the existence of a limiting value of a sequence of numbers $z_1, z_2, z_3, \ldots$ is that corresponding to any given positive number ϵ, however small, it is possible to find a number n such that

$$| z_{n+p} - z_n | < \epsilon$$

for all positive integer values of p.

Cavalieri's Theorem

Two bodies, having equal height and bases, and with all sections parallel to and the same distance from these respective bases also equal, have the same volume.

Cayley-Hamilton Theorem

In matrix theory: If A is a square matrix and $| f(\lambda) | = | A - \lambda I | = 0$ its characteristic equation where I is the unit matrix, then

$$| f(A) | = 0.$$

Cayley-Klein Parameters

Quantities used to express the orientation of a rigid body. If the parameters are represented by a, β, γ and δ, then they are related to the **Eulerian Angles** by

$$a = e^{i(\chi + \phi)/2} \cos \theta/2$$
$$\beta = ie^{i(\chi - \phi)/2} \sin \theta/2$$
$$\gamma = ie^{-i(\chi - \phi)/2} \sin \theta/2$$
$$\delta = e^{-i(\chi + \phi)/2} \cos \theta/2$$

Cayley Numbers

These have eight base elements, $1, e_1, \ldots e_7$, with

$$e_i{}^2 = -1 \qquad e_i\, e_j = -e_j\, e_i$$

$$\text{and } e_1\, e_2 = e_3 \qquad e_1\, e_4 = e_5 \qquad e_1\, e_6 = e_7$$

$$e_2\, e_4 = -e_6 \qquad e_3\, e_4 = e_7 \qquad e_3\, e_5 = e_6.$$

All other products of base elements are obtained by the further rule that $e_i\, e_k = e_j$ implies that $e_k\, e_j = e_i$ and $e_j\, e_i = e_k$.

Celsius Scale
The original Celsius scale of temperature of 1742 has as zero the boiling point of water and as $100°$ the freezing point. The scale was inverted by Christin in 1743. The terms Celsius and Centigrade are used synonymously.

Cerenkov Effect *See* Cherenkov Effect

Césaro's Summation Formula
A method of obtaining a sum of a series Σx_i which may not necessarily be convergent. If $s_1, s_2, \ldots s_n$ are partial sums of the series, then the sum is given by

$$\lim_{n \to \infty} \frac{s_1 + s_2 + \ldots s_n}{n}.$$

For example, for the series $1 - 1 + 1 - 1 + \ldots$, the method obtains a value of $\frac{1}{2}$.

Ceva's Theorem
If three concurrent straight lines are drawn from the vertices A, B and C of a triangle to meet the opposite sides, produced if necessary, at P, Q and R respectively, then the product $\overline{AR} \cdot \overline{BP} \cdot \overline{CQ}$ of three alternate segments taken in order is equal to the product $\overline{RB} \cdot \overline{PC} \cdot \overline{QA}$ of the other three.

Chapman Equation
The viscosity of a gas is equal to:

$$\frac{0{\cdot}499\, mv}{2\pi d^2\, (1 + C/T)}$$

where m is the molecular mass, v its average velocity, d the collision diameter of the molecule.

C is a constant (the Sutherland Constant).

Charles' Law

At a constant pressure, the volume of a definite mass of an ideal gas is proportional to its temperature on the absolute scale. (*See also* **Boyle's Law.**)

Charlier's Checks

A method of checking the arithmetic involved in calculating the mean value x and standard deviation for a number of observations x_s of frequency f_s. It uses

$$\Sigma f_s (d_s + 1) = \Sigma f_s d_s + \Sigma f_s$$

$$\Sigma f_s (d_s + 1)^2 = \Sigma f_s d_s^2 + 2 \Sigma f_s d_s + \Sigma f_s$$

where d_s is the deviation of x_s from a working mean m.
[Using this notation

$$x = \frac{\Sigma f_s d_s}{\Sigma f_s} + m \text{ and } \sigma^2 = \frac{\Sigma f_s d_s^2}{\Sigma f_s} - (m - x)^2]$$

Charpy Test

In metallurgy, an impact test in which a notched specimen, fixed at both ends, is hit by a pendulum.

Chartier's Test

If $f(x) \to 0$ steadily as $x \to \infty$ and if $\left| \int_a^x \phi(x) \, dx \right|$ is bounded as $x \to \infty$

then $\int_a^\infty f(x) \phi(x) \, dx$ is convergent.

Chebyshev Polynomials

These are functions

$$T_n(x) = \frac{1}{n} \sqrt{\left(\frac{2}{\pi}\right)} \cosh(n \cosh^{-1} x)$$

$$U_n(x) = \frac{1}{\sqrt{(x^2-1)}} \sqrt{\left(\frac{2}{\pi}\right)} \sinh\left\{(n+1)\cosh^{-1} x\right\}$$

called Chebyshev polynomials of the first and second kind respectively. The polynomial form arises on substituting $z = \cosh x$ into these functions. They are special cases of the **Gegenbauer Polynomials** and are solutions of the differential equation

$$(x^2 - 1)\frac{d^2y}{dx^2} + x\frac{dy}{dx} - n^2y = 0$$

Chebyshev's Approximation

An arbitrary function can be expanded in a given finite range as a power series (e.g. Taylor's Series). An expansion in terms of **Chebyshev Polynomials** will ensure that the value of the deviation of the resulting polynomial from the original function will be not greater than a given pre-assigned number.

Chebyshev's Inequality

If x is a random variable of mean value $\overline{x}$ and σ is the standard deviation, then there is an upper limit to the probability that the absolute deviation of x from $\overline{x}$ is greater than or equal to a given value a (> 0), and this probability limit is given by

$$P\left[\,|x - \overline{x}| \geqslant a\,\right] \leqslant \frac{\sigma^2}{a^2}$$

Cherenkov Radiation

Radiation which arises when a charged particle traverses a medium at a velocity greater than the phase velocity of light in that medium. The particle sets up local polarization of the dielectric medium and as the coherent growth and decay of a sequence of dipoles occurs along the path of the particle, radiation is excited at an angle θ to the path where $\cos\theta = c/vn$. v is the particle velocity, c is the velocity of light *in vacuo* and n is the refractive index of the medium. Thus, the direction of radiation lies in the surface of a cone of semi-angle θ, and the electric vector of the radiation is perpendicular to the surface of this cone.

Child's Law

In a diode, if certain assumptions are made

$$I_a = KE_a{}^\eta.$$

where I_a and E_a are the anode current and voltage respectively and η and K are constants. For two parallel plane electrodes or for long coaxial cylinders $\eta = 1\cdot5$.

Chladni's Figures
Patterns set up by sprinkling sand upon a glass plate which is set vibrating with a bow. The plate is supported centrally and different symmetrical patterns are obtained by gripping one edge of the plate at different points which thereby become nodes.

Christoffel Symbols
In three dimensions, the Christoffel Symbols are three index symbols defined as follows (for orthogonal co-ordinates).

$$\begin{bmatrix} i \\ i\,i \end{bmatrix} = \frac{1}{h_i}\frac{\partial h_i}{\partial \xi_i}, \quad \begin{bmatrix} i \\ i\,j \end{bmatrix} = \begin{bmatrix} i \\ j\,i \end{bmatrix} = \frac{1}{h_i}\frac{\partial h_i}{\partial \xi_j}$$

$$\begin{bmatrix} j \\ i\,i \end{bmatrix} = -\frac{h_i}{h_j^2}\frac{\partial h_i}{\partial \xi_j}, \quad \begin{bmatrix} i \\ j\,k \end{bmatrix} = 0 \text{ for } i, j, k \text{ all different}$$

where ξ_i, ξ_j, ξ_k are those generalized co-ordinates, h_i, h_j, h_k are corresponding scale factors, and i, j, k can take the values 1, 2, 3 in this succession.

They are not tensors and are used in the formation of derivatives of vectors such as divergence and curl from one co-ordinate system to another.

$$\Gamma^t_{rs} = \sum_{a=1}^{n} \left\{ g^{ta}\, \Gamma_{rs,a} \right\}$$

where the g^{rs} denote the elements of the reciprocal matrix to the matrix g_{rs}. Sometimes Γ^t_{rs} is written $\begin{Bmatrix} r\ s \\ t \end{Bmatrix}$. The four index symbols are defined by equations of the type

$$R^q_{p,mt} = \frac{\partial}{\partial x_i}\,\Gamma^q_{pm} - \frac{\partial}{\partial x_m}\,\Gamma^q_{tp} + \sum_{a=1}^{n} \left\{ \Gamma^a_{pm}\,\Gamma^q_{at} - \Gamma^a_{pt}\,\Gamma^q_{am} \right\}$$

Sometimes $R^q_{p,mt}$ is written $\{pq,\,mt\}$.

Associated with these is a second set of four-index symbols $R_{pq,mt}$ or $[pq,\,mt]$ defined by the equations

$$R_{pq,mt} = \sum_{a=1}^{n} \left\{ g_{qa}\,R^a_{p,mt} \right\}.$$

Chugaev Reaction
S-methyl-O alkyl xanthates containing $\alpha\beta$ hydrogen to the alkyl group pyrolyse to yield olefines without rearrangement of the carbon skeleton.

$$H(CR_2)_2\ O.CS.SCH_3 \rightarrow R_2C = R_2C + CH_3SH + COS$$

Clairant's Theorem
If g_e and g_λ are the values of the acceleration due to gravity at the earth's equator and latitude λ respectively, and m is the ratio of the centrifugal acceleration to g_e, then

$$g_\lambda = g_e\left[1 + (5/2\,m - e)\sin^2\lambda\right],$$

where e = ellipticity = $(r_1 - r_2)/r_1$ and r_1 and r_2 are the equatorial and polar radii. Bouguer and Faye have given correction rules for local variation of altitude.

Clairaut's Differential Equation
If

$$y = x\frac{dy}{dx} + f\left(\frac{dy}{dx}\right)$$

the solution is $y = cx + f(c)$.

The singular solution is obtained by eliminating dy/dx between

$$y = x\frac{dy}{dx} + f\left(\frac{dy}{dx}\right)$$

and

$$x + \frac{d}{dx} f\left(\frac{dy}{dx}\right) = 0$$

Claisen Condensation

A ketoester may be formed by the reaction of two esters having a-hydrogen atoms in the presence of sodium ethoxide, sodamide or triphenyl methyl sodium. In certain cases the second ester does not contain an a-hydrogen atom, viz. ethyl benzoate, ethoxide, sodamide or ethyl formate. The condensation also occurs between esters and ketones to give a $1:3$ diketone. The best known use of this reaction is the preparation of aceto-acetic ester when ethyl acetate condenses with itself in the presence of sodium ethoxide.

$$2CH_3.CO.OC_2H_5 \xrightarrow{NaOC_2H_5} CH_3.CO.CH_2.COO.C_2H_5 + C_2H_5OH$$

Claisen Rearrangement

When heated to about $200°C$, allyl ethers of phenols rearrange to form the corresponding o-allylphenols, e.g.

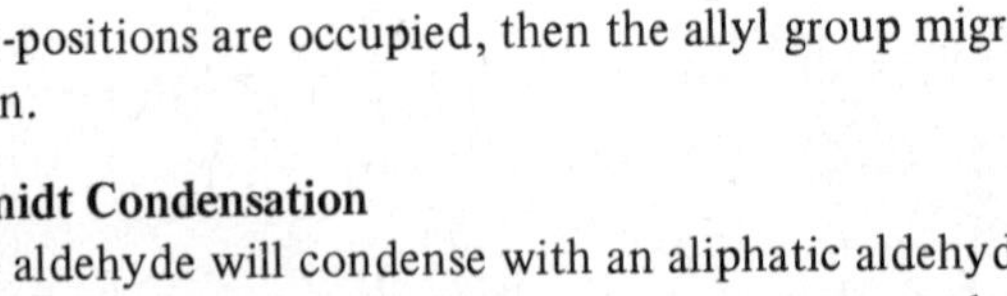

If both the o-positions are occupied, then the allyl group migrates to the p-position.

Claisen-Schmidt Condensation

An aromatic aldehyde will condense with an aliphatic aldehyde or a ketone in the presence of dilute alkali to give an unsaturated aldehyde.

$$C_6H_5CHO + CH_3CHO \xrightarrow{NaOH} C_6H_5CH:CHCHO + H_2O$$

Clapeyron Equation of State

The combined gas equation for one gm-molecule is,

$$pV = RT$$

62

In order to apply this to an arbitrary quantity of gas the number n of moles it contains must be known. The equation can then be written in the form

$$pV = nRT$$

where $R = 8.313 \times 10^7$ ergs degree^{-1} mole^{-1}. This equation is known as the Clapeyron Equation of State for Gases.

Clapeyron's Theorem
If s_i are the displacement components and F_i are the corresponding force components when a body is deformed, where $i = 1, 2, 3, \ldots$ then the strain energy is given by

$$U = \tfrac{1}{2} \Sigma F_i s_i$$

(The F_i's include all deforming loads and body forces but not the six constraints required to hold the body in equilibrium by preventing translation and rotation.)

Clark Process
Softening of water by the addition of lime water to convert acid carbonates to normal carbonates.

Claude Process
The liquefaction of air in stages, the expanding gas being cooled by performing work on pistons.

Claus Benzene Formula *See* Kekulé Benzene Formula

Clausius-Clapeyron Equation
A thermodynamic equation of the form

$$\frac{dp}{dT} = \frac{\Delta H}{T \Delta V}$$

giving the rate of change of pressure with temperature in terms of the increase of heat content in a system consisting of liquid and vapour in equilibrium and the increase of volume accompanying vaporization.

63

Clausius Equation

A modified gas equation of the form:

$$\left\{ p + \frac{a}{T(V+c)^2} \right\}(V-b) = RT$$

where a, b and c are constants.

Clausius-Mosotti Equation

An expression for molar polarization in terms of the dielectric constant, ϵ_r, the molecular weight, M, and the density, ρ.

$$P = \frac{\epsilon_r - 1}{\epsilon_r + 2} \cdot \frac{M}{\rho} \, .$$

If in the optical range ϵ_r is replaced by the square of the complex refractive index, n^{*2}, the **Clausius-Mosotti-Lorentz-Lorenz Equation** is obtained. (*See also* **Lorentz-Lorenz Equation**.)

Clausius' Statement (of the Second Law of Thermodynamics)

No process is possible whose sole result is the removal of heat from a reservoir at one temperature and the adsorption of an equal quantity of heat by a reservoir at a higher temperature. (*See also* **Kelvin's Statement**.)

Clausius' Theorem

In an arbitrary heat cycle

$$\oint \frac{dQ}{T} \leqslant 0$$

where dQ is an infinitesimal heat adsorption. The equality applies to a reversible cycle and the inequality to an irreversible cycle.

Clausius' Virial Theorem

For a system of n particles having position vectors r_i ($i = 1, 2, \ldots n$) and acted upon by forces F_i, the average kinetic energy $\overline{T}$ of all the particles averaged over time is

$$\overline{T} = -\tfrac{1}{2} \overline{\sum_i F_i \cdot r_i}$$

Clebsch-Gordan Coefficients
If D_i and D_k are irreducible representations (see **Schur's Lemma**) and

$$D_i \oplus D_j = \sum_k C_{ij}^k D_k$$

then C_{ij}^k are Clebsch-Gordan coefficients and the above expression, called the Clebsch-Gordan series, gives the number of times the irreducible representation D_k occurs in the **Kronecker Product** of D_i and D_j. The series enables the determination of selection rules for transitions between energy levels.

Clemmensen Reduction
A carbonyl group may be reduced to a methylene group by zinc amalgam and concentrated hydrochloric acid. This reaction works better with ketones than with aldehydes.

Coehn's Rule
In electrification by friction the material with the higher dielectric constant becomes positively charged. The charge density in coulombs m^{-2} is 15×10^{-6} $(\epsilon_1 - \epsilon_2)$ where ϵ_1 and ϵ_2 are the permittivities of the materials.

Cole and Adie's Method
A colorimetric method for determining the pH of the gastric juices using a thymol-blue indicator.

Cole's Ferricyanide-Methylene Blue Method (for the Estimation of Reducing Sugars)
Potassium ferricyanide is treated with alkali and boiled. The mixture is titrated with the sugar solution. When the yellow colour has nearly disappeared, a drop of methylene blue is added. The titration is continued until the colour is discharged. Both methylene blue and ferricyanide are reduced by these sugars but methylene blue is not affected until all the ferricyanide has been reduced.

Cole's Method for Acidity of Urine
A colorimetric method using a phenol red indicator.

Cole's Method for Amino Acids and Ammonia in Urine

When neutral ammonium salts or amino acids are treated with formaldehyde, hexamethylene tetramine is formed and hydrochloric acid is liberated.

$$4NH_4Cl + 6HCHO \rightarrow (CH_2)_6N_4 + 6H_2O + 4HCl$$

The amount of alkali required to make the solution neutral again allows the estimation of the amount of ammonia present. This estimation is carried out using the Cole-Onslow comparator.

Cole's Test for Bile Pigments

About 15c.c. of fluid are boiled and two drops of saturated magnesium carbonate are added. 10% barium chloride is added until no further precipitate is obtained. Boil, allow to stand, decant, filter, transfer the precipitate to a dry test tube. Add 3c.c. of absolute alcohol and shake, then add 3 drops of conc. sulphuric acid and 1 drop 2% potassium chlorate solution. Boil for ½ minute. A green coloration in the alcohol indicates bile pigments.

Compton Effect

Important evidence for the particle or photon nature of radiation is provided by the discovery by A. H. Compton that if monochromatic x-rays are allowed to fall on a scattering material of low atomic weight, e.g. carbon, the scattered x-rays contain, in addition to those having the incident wavelength, others of somewhat longer wavelength. The change of wavelength is given by

$$\Delta\lambda = \frac{h}{mc}(1 - \cos\theta)$$

where θ is the angle of scattering of a photon and m is the rest mass of an electron. If E_p is the energy of the incident photon and E_p' the energy of the scattered photon, then

$$E_p' = E_p\left[1 + \frac{E_p}{mc^2}(1 - \cos\theta)\right]^{-1}$$

Compton-Getting Effect

The sidereal variation of cosmic radiation of extra-galactic origin due to rotation of the earth's galaxy.

Compton Wavelength *See* **Appendix**

Conwell-Weisskopf Equation
The scattering of electrons by singly ionized donors or acceptors in semiconductors has been considered by these two workers. They have found tnat the mobility, μ, of the electron is given by

$$\mu = \frac{2^{7/2}\, \epsilon_r^2\, (kT)^{3/2}}{N_e\, \pi^{3/2}\, e^3\, m^{*\,1/2} \ln(1 + x^2)}$$

where

$$x = \frac{6\, \epsilon_r\, dkT}{e^2}$$

N_e is the concentration of ionized donors, $2d$ is the average distance between near ionized donor neighbours, m^* is the effective mass of the electron and ϵ_r is the dielectric constant.

Cooper Pair
In superconductivity theory it has been shown that electrons with energy close to the **Fermi energy** can form quasi-molecular pairs called Cooper pairs which can then behave as bose particles subject to **Bose-Einstein Statistics**.

Corbino Disc
A thin cylindrical disc perforated with a central hole. It is used for making electrical measurements in a magnetic field with the current flowing radially and the magnetic field perpendicular to the plane of the disc. In this arrangement the **Hall Effect** is shorted out and there is a maximum magnetoresistance called **Corbino magnetoresistance**.

Coriolis' Force
Owing to the earth's rotation a body moving freely upon the earth is deflected to the right in the northern hemisphere and the left in the southern hemisphere. If the velocity of motion is v then (when v is not too great) the component of acceleration β at right angles to the direction of motion is

$$\beta = \frac{4\pi v \sin\phi}{T}$$

where ϕ is the geographical latitude and T is the period of the earth's rotation. According to **Newton's Laws**, each acceleration may be explained by means of a force and in this sense we can speak of 'the deflecting force of the earth's rotation'. Attention to this force was first drawn by G. G. Coriolis and these forces are sometimes known by his name.

Cornu-Hartman Formula

A relation between the deviation D produced by a prism and the wavelength λ of the light passing through it

$$\lambda = \lambda_0 + \frac{C}{D - D_0}$$

where λ_0, C and D_0 are constants.

Cornu's Spiral

A double spiral used in computing intensities in a Fresnel diffraction pattern, obtained by plotting

$$x = \int_0^v \cos \frac{\pi v^2}{2} \, dv \text{ against } y = \int_0^v \sin \frac{\pi v^2}{2} \, dv$$

Cotes' Rule

For obtaining the area under a curve, assume for y

$$y = A_0 + A_1 x + A_2 x^2 + \ldots + A_{n-1} x^{n-1}$$

Determine the coefficients $A_1, A_2, \ldots, A_{n-1}$ so that, for the n equidistant values of x, y shall have the prescribed values $y_1, y_2, \ldots, y_n$. The area is then given by

$$\int y \, dx = A_0 x + \frac{1}{2} A_1 x^2 + \frac{1}{3} A_2 x^3 + \ldots + \frac{1}{n} A_{n-1} x^n$$

taken between proper limits of x.

Cotton-Mouton Effect

When light is passed through certain pure liquids in a direction which is perpendicular to an applied magnetic field, the liquids show double refraction which is much greater than would appear due to the **Voigt**

Effect. It arises when the molecules of a liquid possess magnetic and optical anisotropy so that when a magnetic field is applied the molecules become aligned. It is proportional to the square of the magnetic field strength and is an analogous effect to the **Kerr** (electro-optic) **Effect.**

Coulomb *See* **Appendix**

Coulomb's Law of Force
The force between two electric charges (or magnetic poles) is proportional to their magnitudes and inversely proportional to the square of the distance between them.

Coulomb's Law of Friction
The force required just to move an object along a plane is given by μp, where p is the pressure normal to the plane and μ is the coefficient of friction which is independent of the area of the surfaces in contact.

Crafts' Rule
A rule based on the **Clausius-Clapeyron Equation** for the correction of boiling points of liquids to atmospheric pressure. If $T\,^{\circ}C$ is the observed boiling point at p mm pressure, then ΔT, the temperature difference to be added to give the boiling point at 760 mm, is

$$\Delta T = c\,(273 + T)\,(760 - p)$$

where $c = 0 \cdot 00012$ for most liquids.

Cramer's Rule
If, in the equations

$$a_{11}x_1 + \ldots + a_{1n}x_n = k_1$$
$$\cdots \cdots \cdots \cdots \cdots$$
$$a_{n1}x_1 + \ldots + a_{nn}x_n = k_n$$

the determinant

$$\Delta = \begin{vmatrix} a_{11} \ldots a_{1n} \\ \cdots \cdots \cdots \\ a_{n1} \ldots a_{nn} \end{vmatrix} \neq 0$$

then the solution of the equations is

$$x_1 = \left| \begin{array}{cccc} k_1 & a_{12} & \cdots & a_{1n} \\ \cdot & \cdot & \cdot & \cdot \\ k_n & a_{n2} & \cdots & a_{nn} \end{array} \right| \quad \cdots x_n = \left| \begin{array}{ccccc} a_{11} & a_{12} & \cdots & a_{1n-1} & k_1 \\ \cdot & \cdot & \cdots & \cdot & \cdot \\ a_{n1} & a_{n2} & \cdots & a_{nn-1} & k_n \end{array} \right|$$
$$ \Delta \Delta$$

Criegee Reaction

The bond between two carboxylated carbon atoms can easily be split by oxidation. If permanganate or dichromate is used two molecules of acid are produced. In anhydroxy solution, lead tetra acetate can be used to stop the reaction at the aldehyde stage; this reaction is knows as the Criegee reaction.

$$\begin{array}{c} R(CHOH) \\ | \\ R(CHOH) \end{array} \xrightarrow{\;Pb(OCOCH_3)_4\;} 2RCHO + 2HOCOCH_3 + Pb\,(OCOCH_3)_2$$

Periodic acid is the preferred reagent in aqueous solution.

Crookes Dark Space

As the gas pressure in a discharge tube is gradually diminished it is found that the glow surrounding the cathode detaches itself leaving a space around the electrode. This gap is known as the Crookes Dark Space. At sufficiently low pressures it fills the whole tube. The Crookes dark space is also known as the **Hittorf Dark Space**. (*See* **Faraday Dark Space**.)

Crum-Brown and Gibson Rule

If a compound HX can be directly oxidized to HXA it is *meta* orientating when substituted in the benzene ring whereas if it cannot be oxidized it will be *ortho-para* orientating. Thus, for example, the nitro group is *meta*-directing as nitrous acid is directly oxidized to nitric acid whereas the hydroxyl group is *ortho-para* directing as water cannot be directly oxidized to hydrogen peroxide.

Crum-Brown and Walker Synthesis

This is a general method of preparation of the even members of the series $HOOC(CH_2)_nCOOH$. If the potassium alkyl esters of dicarboxylic acids are electrolysed the following reaction occurs

$$2(C_2H_5OOC(CH_2)_nCOO)' \rightarrow \begin{matrix} C_2H_5OOC(CH_2)_n \\ | \\ C_2H_5OOC(CH_2)_n \end{matrix} + 2CO_2$$

Curie *See* Appendix

Curie Point (Temperature)

Ferromagnetic materials lose their permanent or spontaneous magnetism when heated above a critical temperature called the Curie point. The value of this critical temperature depends upon the material. Ferroelectric materials likewise lose their spontaneous polarization above an analogous critical temperature. This upper Curie point must be clearly distinguished from the lower Curie point below which temperature the property of ferroelectricity itself disappears.

Curie Principle

In the thermodynamics of irreversible processes, the existence of spatial symmetry properties in a system may simplify the form of the phenomenological equations in such a way that the Cartesian components of the fluxes do not depend on all the Cartesian components of the thermodynamic forces. (*See also* **Onsager's Reciprocal Relations.**)

Curie's Law

In a paramagnetic substance

$$M = \frac{C}{T} H$$

where M is the intensity of magnetization, H is the magnetic field strength and C is a constant, independent of the absolute temperature T, called the Curie constant (*see* **Curie-Weiss Law**).

Curie-Weiss Law

In a ferromagnetic substance the magnetic susceptibility above the **Curie Temperature** is given by

$$\chi_m = \frac{C}{T - T_C}$$

where C is the Curie constant $= N\mu^2/3k$ and T_C is the Curie temperature.

71

N is the number of atoms per unit volume having magnetic moment μ. The law can also be applied to electric susceptibility in ferroelectrics.

Curtius Rearrangement

Acyl azides are prepared by the reaction of an acyl chloride on sodium azide.

$$RCOCl + NaN_3 \rightarrow RCON_3 + NaCl$$

These acyl azides undergo rearrangement when heated to produce an isocyanate (*see* **Hofmann Rearrangement**).

Czockralski Method

Method of single crystal growth in which a seed crystal (or alternatively, if this is not available, a wire or small capillary tube) is dipped into a melt of the material and then very slowly withdrawn. By keeping the seed crystal below the temperature of the melt, more material solidifies on to it and a large single crystal is gradually grown.

D

D'Alembertian
A differential operator

$$\square^2 V = \frac{\partial^2 V}{\partial x^2} + \frac{\partial^2 V}{\partial y^2} + \frac{\partial^2 V}{\partial z^2} - \frac{1}{c^2}\frac{\partial^2 V}{\partial t^2}$$

Thus $\square^2 V = 0$ is the wave equation for waves travelling with a velocity c.

D'Alembert's Differential Equation *See* Lagrange's Differential Equation

D'Alembert's Paradox
A body of any shape completely immersed in an incompressible non-viscous fluid which is steadily moving experiences no drag.

D'Alembert's Principle
For a system of particles of mass m_i acted upon by the real forces F_i subject to constraints such that at a certain instant they have acceleration a_i then

$$F_i - ma_i = 0 \text{ for all } i.$$

ma_i is sometimes called the 'effective force' acting on the particle, in which case D'Alembert's Principle can be stated as: the reversed effective forces and the real forces together give statical equilibrium.

D'Alembert's Test for Convergence

In $$\Sigma\, u_n = u_1 + u_2 + \ldots$$

if for all n's greater than r

$$\left| \frac{u_n + 1}{u_n} \right| < \rho$$

where ρ is a positive number less than unity and independent of n, $\Sigma\, u_n$ is absolutely convergent.

Dalton's Law of Evaporation
Although the pressure of a gas over a liquid decreases the rate of evaporation, the partial pressure of the vapour at equilibrium is independent of the presence of other gases and vapours.

Dalton's Law of Partial Pressure
The total pressure of a mixture of gases is equal to the sum of the partial pressures of the constituent gases. The partial pressure is defined as the pressure each gas would exert if it alone occupied the volume of the mixture at the same temperature.

Dalton's Law of Solubility of Gases
The individual constituents of a gaseous mixture are dissolved in the ratio of their partial pressure, that is, independently of one another.

Dalton's Temperature Scale
In the Kelvin scale the unit of temperature measurement is always the same and is given by

$$t = T - 213 \cdot 16$$

where t is the temperature on the Centigrade scale and T is the temperature on the Kelvin scale.

On the Dalton scale the degree is defined by the relation

$$t = 273 \left\{ \left(\frac{373}{273} \right)^{\tau/100} - 1 \right\}$$

where τ is the temperature on the Dalton scale. Thus on the Kelvin scale absolute zero is at $273 \cdot 16$ and on the Dalton scale it is at $-\infty^{\circ}\text{C}$.

The concept of absolute zero only arises because of the method of definition on the Kelvin scale and has no physical significance. It happens, however, that the fundamental laws are much simpler when expressed on this scale than on the Dalton scale.

Darzen's Glycidic Ester Condensation
A reaction involving the condensation of an aldehyde or ketone with an a-haloester in the presence of sodium ethoxide, sodamide or potassium t-butoxide to give an a-β-epoxy ester.

$$C_6H_5 . COCH_3 + ClCH_2 . COOC_2H_5 \xrightarrow{\text{NaNH}_2} \underset{CH_3 \quad O}{\overset{C_6H_5}{C} - CH . COOC_2H_5}$$

These glycidic esters are of interest for, on hydrolysis and decarboxylation, an aldehyde or ketone is produced with a higher carbon content than the original aldehyde or ketone used. Chloracetic ester yields an aldehyde whilst substituted chloracetic esters give rise to ketones.

Darzen's Procedure
Alkyl halides may be prepared by refluxing one molecule of an alcohol with one molecule thionyl chloride in the presence of one molecule pyridine.

$$ROH + SOCl_2 \xrightarrow[\text{pyridine}]{} RCl + SO_2 + HCl$$

Dauphiné Twin
A type of twinning which occurs in crystals of quartz. The twin plane is the $(5, 1, 2)$ plane (*see* **Miller Index**).

Deacon's Process
This process produces chlorine very cheaply. A mixture of air and hydrochloric acid gas is passed over a catalyst, usually cupric chloride, at 400-450°C.

$$4HCl + O_2 \rightleftharpoons 2H_2O + 2Cl_2 - 14 \text{ K. cals.}$$

The yield is about 60% and the excess hydrochloric acid is dissolved out of the gas. The chlorine so produced contains a large proportion of nitrogen and is used mainly for the preparation of bleaching powder.

De Broglie's Theory
The dualism observed between the wave and particle functions of radiation led De Broglie to postulate that a similar dualism might exist in material particles such as electrons. It can be shown that the wavelength λ of the hypothetical matter waves can be represented by the equation

$$\lambda = \frac{h}{mv} = \frac{k}{p}$$

where v is the group velocity of the matter waves. (The product of group velocity and the normal phase velocity is equal to the square of the velocity of light.) m is the mass of the particle and thus mv is equal to the momentum p.

De Brun-van Eckstein Rearrangement
If, for example, glucose is mixed with aqueous calcium hydroxide and allowed to stand, it isomerizes to produce a mixture of glucose, fructose, mannose, unfermented ketoses and small proportions of other products. Other aldoses and ketoses behave similarly. This rearrangement has been used to prepare certain ketoses.

Debye *See* **Appendix**

Debye Equation (for heat capacity of a solid)
Debye, from a quantum mechanical calculation of the energy of a coupled system of $3N$ oscillators, obtained an equation for the variation of the heat capacity C_V of a solid, with temperature, of the form

$$C_V = \frac{9R}{v_m{}^3} \int_0^{v_m} \frac{\exp hv/kT}{(\exp hv/kT - 1)^2} \left(\frac{hv}{kT} \right)^2 v^2 \, dv$$

where v_m is the **Debye frequency**.
The quantity hv/k has the dimensions of temperature and the special value hv_m/k, where v_m represents the maximum frequency of the $3N$ vibration frequencies for each substance, is called its characteristic temperature θ. Also hv/kT is dimensionless and may be replaced by the variable x. Then

$$C_V = 3R \left(\frac{12T^3}{\theta^3} \int_0^{\theta/T} \frac{x^3 \, dx}{\exp x - 1} - \frac{3\theta/T}{\exp \theta/T - 1} \right).$$

A method has been devised for evaluating the integral and the results have been tabulated for various values of θ/T. In order to determine the heat capacity of an element at any temperature it is only necessary to know the value at one temperature so that θ may be determined and then by use of these tables the value of C_V at any other temperature may be calculated. Good agreement with experimental results has been obtained.

Debye Equation for Polarization
The total polarization P for N molecules was calculated by P. Debye.

in terms of the permanent moment of the molecule μ and its polarizability a;

$$P = \tfrac{4}{3}\pi Na + \tfrac{4}{3}\pi N\left(\frac{\mu^2}{3kT}\right)$$

Debye-Hückel Equation

Until 1923 the activity coefficient was regarded as an empirical quantity. By taking into account the interionic attraction Debye and Huckel found it possible to calculate the ratio of activity to concentration of an ion in dilute solution. The equation reduces for a given solvent at a definite temperature to:

$$-\log f_i = A\, Z_i^2\, \sqrt{I}$$

where f_i is the activity coefficient of the ion, Z_i is the valency of the ion, I is the ionic strength ($\tfrac{1}{2}C_i Z_i^2$ where C_i is the number of gram ions per litre) and

$$A = \frac{2N^2 e^3 \pi / 1000}{2\cdot303 R^{3/2}(\epsilon_r T)^{3/2}}$$

ϵ_r is the dielectric constant of the solvent.

Debye-Hückel-Onsager Equation *See* Onsager Conductivity Equation

Debye-Jauncey Scattering

Incoherent background scatter occurring when x-rays are scattered from a crystal. The intensity is low compared with the Bragg reflections (*see* **Bragg's Law**) at low temperatures but increases with increase of temperature.

Debye-Scherrer Method (Powder Method)

An x-ray diffraction method in which the sample is used in the form of a powder. The sample, contained in a thin-walled silica tube or stuck to a thin fibre, is rotated in a beam of monochromatic x-radiation. The diffraction pattern is usually recorded on a cylindrical film whose axis is parallel to the axis of the specimen. Under these conditions the pattern consists of a series of concentric rings whose spacings obey **Bragg's Law** and can be used to obtain the lattice parameters of the sample.

77

Debye-Sears Effect

A piezo-electric crystal vibrating in a liquid sets up a train of acoustic waves whose nodes occur every half wavelength. It is thus possible to obtain the effect of a diffraction grating by passing a beam of light through the liquid in a direction at right angles to the wave direction.

Debye T^3 Law

At low temperatures, the heat capacity of a crystalline solid is approximately proportional to T^3. This can be shown from the **Debye Equation**.

Debye-Waller Factor

When x-rays are diffracted from a crystal lattice the intensity is reduced due to thermal motion of the atoms by the Debye-Waller factor

$$\exp(-1/3 <u^2><\Delta k^2>)$$

where $<u^2>$ is the mean square displacement of the atoms from their equilibrium position and Δk is the change in wave vector of the x-rays on reflection.

Dedekind-Peirce Theorem

A class S is infinite if, and only if, it can be put into one to one correspondence with a proper subset of itself.

Dedekind's Definition

Suppose a certain rule divides a whole system of rationals into two classes, a lower class A and an upper class A', so that any number in A is less than any number of A'. Either

(1) A has a greatest number or A' has a least number, in which case the classification defines an irrational number which is to follow all in A or the least in A', or

(2) if there is no greatest number in A and no least number in A', the classification defines an irrational number which is to follow all the numbers in A and to precede all in A'.

Dedekind's Test

If Σa_n has bounded partial sums, $\Sigma |b_n - b_{n+1}|$ is convergent and $b_n \to 0$, then $\Sigma a_n b_n$ is convergent.

De Gua's Rule

If a group of r consecutive terms is missing from a polynomial $f(x) = 0$ then
   (1) if r is even, the equation has at least r imaginary roots,
   (2) if r is odd, there are at least $r + 1$ or at least $r - 1$ imaginary roots according as the terms which immediately precede and follow the group have like or unlike signs.

De Haas–van Alphen Effect

Oscillations of the diamagnetic susceptibility of a metal as a function of a strong magnetic field. The effect arises from the quantization of the energy of the electrons and hence the quantization of the electron orbits in the presence of a magnetic field. The latter must be sufficiently strong for electrons to complete several orbits before colliding with imperfections or the surface of the specimen.

Delambre's Analogies

In the solution of spherical triangles, if a, b and c are the sides of the triangle and a, β and γ are the angles opposite a, b and c respectively, then

$$\sin \tfrac{1}{2}a \cos \tfrac{1}{2}(b + c) = \cos \tfrac{1}{2}a \cos \tfrac{1}{2}(\beta + \gamma)$$
$$\cos \tfrac{1}{2}a \sin \tfrac{1}{2}(b - c) = \sin \tfrac{1}{2}a \sin \tfrac{1}{2}(\beta - \gamma)$$
$$\sin \tfrac{1}{2}a \sin \tfrac{1}{2}(b + c) = \sin \tfrac{1}{2}a \cos \tfrac{1}{2}(\beta - \gamma)$$
$$\cos \tfrac{1}{2}a \cos \tfrac{1}{2}(b - c) = \cos \tfrac{1}{2}a \sin\tfrac{1}{2}(\beta + \gamma)$$

Also **Napier's Analogies:**

$$\tan \tfrac{1}{2}(b + c) \cos \tfrac{1}{2}(\beta + \gamma) = \tan \tfrac{1}{2}a \cos \tfrac{1}{2}(\beta - \gamma)$$
$$\tan \tfrac{1}{2}(\beta + \gamma) \cos \tfrac{1}{2}(b + c) = \cot \tfrac{1}{2}a \cos \tfrac{1}{2}(b - c)$$
$$\tan \tfrac{1}{2}(b - c) \sin \tfrac{1}{2}(\beta + \gamma) = \tan \tfrac{1}{2}a \sin \tfrac{1}{2}(\beta - \gamma)$$
$$\tan \tfrac{1}{2}(\beta - \gamma) \sin \tfrac{1}{2}(b + c) = \cot \tfrac{1}{2}a \sin \tfrac{1}{2}(b - c)$$

See also **Napier's Rules.**

De la Rue and Miller's Law

The sparking potential V of a gas at pressure p is a function of pd only, where d is the sparking distance, for a field between two parallel plates. The equation connecting d, p and V is of the form

$$d = \ln f\left(\frac{V}{pd}\right) - \ln \phi\left(\frac{V}{pd}\right)$$

Delbrück Scattering
Scattering of photons by the Coulombic field of a nucleus.

Delépine Reaction
Reaction of hexamethylene tetramine with primary alkyl halides yields a salt. This compound can be easily hydrolysed to give formaldehyde, ammonia and a primary amine.

$$(CH_2)_6N_4 + RX \rightarrow \left[(CH_2)_6N_4R\right]X$$

$$\left[(CH_2)_6N_4R\right]X + 6H_2O + 3HX \rightarrow RNH_3X + 6HCHO + 3NH_4X$$

Dellinger Fadeout
A failure of short-wave radio communication caused by a highly absorbing D layer in the ionosphere associated with a sunspot.

Dember Effect
The establishment of a voltage in a conductor or semiconductor by illumination of one surface. Hole-electron pairs are produced by the illumination and because the electrons are more mobile than the holes, a negative charge is established on the non-illuminated side. Hence an electric field is set up preventing further flow. It is also called the photo-diffusion effect.

Dem'janov Rearrangement
When, say, cyclopropylmethylamine or cyclobutylamine react with nitrous acid, ring expansion occurs in the first case and ring contraction in the second to give the same mixture of cyclopropylcarbinol and cyclobutanol (along with a small amount of allylcarbinol). Changes in ring size due to the reaction of amines with nitrous acid are known as Dem'janov rearrangements.

De Moivre's Theorem
If n is a positive or negative integer or a fraction

$$(\cos x + j \sin x)^n = \cos nx + j \sin nx$$

De Morgan's Rules
These apply to **Boolean Algebra**. Any logical or binary expression equals the complement (or negation) of the expression obtained by changing all logical products (sometimes called AND statements) to logical sums (sometimes called OR statements) or vice versa and replacing variables with their complements.

$$\overline{A + B} = \overline{A}\,\overline{B} \text{ and } \overline{A\,B} = \overline{A} + \overline{B}$$

where $\overline{A}$ and $\overline{B}$ are the complements of A and B respectively.

De Morgan's Test
A series Σu_n in which $\lim_{n \to \infty} \left| \dfrac{u_{n+1}}{u_n} \right| = 1$ will be absolutely convergent if a

positive number c exists such that $\lim_{n \to \infty} \left\{ \left| \dfrac{u_{n+1}}{u_n} \right| - 1 \right\} = -1 - c$.

Descartes' Folium
A curve having the equation
$$x^3 + y^3 = 3\,a\,x\,y$$

Descartes' Laws
When a beam of light passes between two isotropic media, the incident and refracted rays and the normal to the surface lie in the same plane. The incident and refracted rays lie on opposite sides of the surface, and the sines of their angles of inclination bear a constant ratio depending on the media used.

Descartes' Rule of Signs
No equation can have more positive roots than it has changes of sign from positive to negative and from negative to positive, in the terms of the equation $f(x) = 0$. No equation can have more negative roots than there are changes of sign in $f(-x)$.

81

Destriau Effect
Certain phosphorescent inorganic materials when suspended in a dielectric film may be excited to luminescence if subjected to the action of an alternating electric field. This effect was discovered by G. Destriau in 1947 and such phosphors are termed electroluminescent. Another form of electroluminescence of great technological importance is carrier injection electroluminescence which occurs when a d.c. field is applied to a semiconductor such as SiC or GaAs. It arises because of radiative recombination of the injected carriers.

Dewar Benzene Formula *See* Kekulé Benzene Formula

d'Huilier's Equation
In the solution of spherical triangles

$$\tan\left(\frac{a}{2} - \frac{\epsilon}{4}\right) = +\sqrt{\left[\frac{\tan \frac{1}{2}(s-b)\, \tan \frac{1}{2}(s-c)}{\tan \frac{1}{2}s\, \tan \frac{1}{2}(s-a)}\right]}$$

where ϵ is the spherical excess $(= a + \beta + \gamma - \pi)$ and $s = \frac{1}{2}(a + b + c)$. See **Delambre's Analogies** for the remaining notation.

Dieckmann Reaction
Certain cyclic ketones may be prepared by the Dieckmann reaction, an intramolecular Claisen condensation. Thus adipic, pimelic and suberic esters when treated with sodium give rise to five-, six- or seven-membered cyclic ketones.

Diels-Alder Reaction
A conjugated diene can be added to an ethylenic compound which has a carbonyl group adjacent to the double bond. The adduct is always a six-membered ring the reaction taking place in the 1:4 positions.

Butadiene Maleic anhydride Tetrahydrophthalic anhydride

Dienes may also be made to add on to quinones, dienes with a carbonyl group adjacent to the double bond and even in certain cases to the $1:2$ double bond of another diene.

Diesel Cycle

Diesel was dissatisfied with the low efficiency of the Otto engine and was led to investigate certain contrivances whereby the efficiency of the Carnot cycle may be reached. This investigation failed but led to the invention of the Diesel engine which depends upon the idealized Diesel cycle in the diagram.

Pure air is sucked in isobarically (E→C) and compressed adiabatically (C→D). A valve is opened and oil is forced in under pressure and burns at the temperature of the system. The supply of oil is regulated so that the piston moves forward and the pressure remains constant (D→A) the temperature changing from

$T_D→T_A$. The oil is cut off at A and the gas undergoes adiabatic expansion (A→B). The valve is opened at B and the pressure drops to C. C→E is a scavenging stroke in which useless gas is forced out and the engine becomes ready for a fresh cycle.

Dieterici Equation

The Dieterici equation is a relation between the pressure of a real gas and its volume of the form

$$p = \frac{RT}{V-b} \exp\left\{-\frac{a}{RTV}\right\}$$

where a and b are both constants. At low pressures this equation reduces to the **Van der Waals Equation.**

Diocles, cissoid of

A curve whose equation is

$$y^2(a-x) = x^3$$

or alternatively

$$r = a\left(\frac{1}{\cos\phi} - \cos\phi\right)$$

83

Dirac Delta Function
A function $\delta(t)$ defined as zero when $t \neq 0$ and infinite at $t = 0$, such that

$$\int_{-\infty}^{\infty} \delta(t)\, dt = 1.$$

Dirac Equation
A relativistic wave equation for an electron in an electromagnetic field. The Dirac wave function has four components and the solution requires that, in order that the total angular momentum of the electron be constant, the electron must possess its own intrinsic angular momentum called 'spin'. The existence of the electron spin was deduced empirically by Uhlenbeck and Goudsmith before Dirac's theory was first published.

Dirichlet Condition
In the Fourier expansion of f(x), the assumption that, when x approaches the ends 0 and 2π of the interval, the function f(x) tends to definite limits which are denoted by f(+ 0) and f(2π − 0).

Dirichlet Problem
In potential theory, the determination of a function, harmonic in a region, when its boundary values are given. (See **Neumann Problem**.)

Dirichlet's Test for Convergence
If Σu_n converges or oscillates between finite limits and $a_1, a_2, a_3, \ldots$ is a decreasing sequence of positive terms which tends to zero as a limit, then

$$\Sigma a_n u_n \text{ is convergent.}$$

Dirichlet Theorem
Every arithmetic progression $a, a + b, a + 2b, \ldots$ in which a, b are integers with no common divisor greater than unity, contains an infinity of primes.

Doebner Miller Synthesis
Methylquinoline can be synthesized by this process which is analogous to the **Skraup Synthesis**. It consists of heating analine with paraldehyde

in the presence of hydrochloric acid, The mechanism probably involves the formation of crotonaldehyde ($CH_3CH{=}CHCHO$), $1:4$ addition to aniline, cyclization and oxidation.

Donnan Potential

A solution of an electrolyte consisting of two diffusible ions is separated by a membrane from a salt consisting of the same diffusible ion as one of those on the other side of the membrane and a non-diffusible ion. At equilibrium the concentration of the common ion is unequal on either side of the membrane. This inequality of concentration is maintained across the membrane by a potential known as the Donnan membrane potential. It can be shown thermodynamically that the fraction of the diffusible ion F_D which diffuses across the membrane from the solution containing the two diffusible ions initially at concentration C_1 into the solution containing the non-diffusible ion initially at concentration C_2, is;

$$F_D = \frac{C_1}{C_1 + 2C_2}.$$

It is smaller the larger the concentration of non-diffusible ion.

Donnay-Harker Principle

Generalization of **Bravais' Law** for crystal habit by considering the space group of a crystal rather than just the **Bravais Lattice** type, i.e. an extended lattice generalized from point group symmetry. When no glide planes or screw axes are present, the principle is equivalent to Bravais' law.

Doppler Effect

The observed frequency of radiation from a source possessing a component of motion in the direction of an observer differs from the true value by an amount proportional to the component. If the velocity of the source is towards the observer, the frequency appears raised, if away from the observer, the frequency is lowered. The frequency as recorded by the observer is

$$\nu' = \nu \, \frac{\sqrt{(1 - v^2/c^2)}}{1 - (v/c)\cos\theta}$$

where ν is the frequency of the source which has velocity v. θ is the

85

angle between the direction of motion of the source and the direction of the observer relative to the source. Due to this effect, the outward movement of galaxies leads to a shift of the observed spectra towards the red end (see **Hubble Shift**).

The above expression must be obtained relativistically but approximates to the classical expression

$$\nu' = \frac{\nu}{1 - (v/c)\cos\theta}$$

for $v \ll c$.

Dorn Effect
When particles fall through water a potential difference is set up. This is a form of electrokinetic effect.

Downs Process

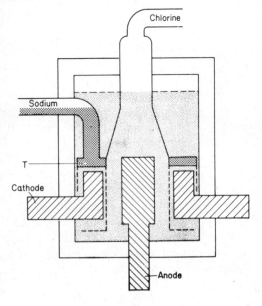

Downs Process

A fused electrolyte of sodium chloride containing potassium chloride and fluoride added to lower the melting point from 800°C to 600°C is electrolyzed using a carbon anode and a ring-shaped cathode. The chlorine released at the anode is led off by means of a funnel over the anode whilst the sodium collects in a ring-shaped inverted trough T and is forced by the pressure of fused sodium chloride into a container. The efficiency is 70% higher than that of the Castner Cell (*see* **Castner Process**).

Draper Effect
At the commencement of the hydrogen-chlorine reaction there is an increase in volume at constant pressure. According to the equation

$$H_2 + Cl_2 \rightarrow 2HCl$$

the volume should be constant. This increase in volume has been traced to a rise in temperature of the reactants due to the heat involved in the reaction.

Draper's Law *See* **Grotthus' Law**

Drude Equation
In its simplest form the Drude equation may be written as

$$[a] = \frac{k}{\lambda^2 - \lambda_0{}^2}$$

where $[a]$ is the specific dispersion, $\lambda_0{}^2$ is the dispersion constant, k the rotation constant and λ is the wavelength of the absorbed light which is usually expressed in microns. Within and close to the absorption band the equation does not hold but in regions of complete transparency an equation involving one or more Drude terms represents in a very satisfactory way the variation of the angle of rotation with the wavelength of the polarized light.

Drude's Theory of Conduction
Drude gave a formula for the specific resistance ρ of a substance, based on classical mechanical considerations, as

$$\frac{1}{\rho} = \frac{1}{2} \frac{Ne^2}{m} t$$

where N is the number of free electrons per c.c., e the charge and m the mass of an electron, and t is the time each free electron moves between two successive collisions with molecules. The formula leads to values of the mean free path of an electron which are far too large and the difficulty becomes much worse for conduction at very low temperatures. Application of quantum mechanics has given a new approach to the subject.

Duane-Hunt Law
The high frequency limit for x-rays is proportional to the voltage applied to accelerate the electron beam used to bombard the metal target.

Du Bois Raymond's Test for Convergence
$\Sigma\, a_s b_s$ is convergent if $\Sigma\, (b_s - b_{s+1})$ be absolutely convergent and if $\Sigma\, a_s$ converge at least conditionally.

Duchemin's Formula
The normal wind pressure N on an inclined surface is given by

$$N = F\,\frac{2 \sin a}{1 + \sin^2 a}$$

where a is the angle of the inclined surface with the horizontal and F is the force of the wind.

Duffing Equation
A differential equation describing the oscillations of a non-linear spring;

$$\frac{d^2 y}{dt^2} + a\frac{dy}{dt} + y + by^3 = 0.$$

When $a > 0$ and $b > 0$ ('hard spring') the solution is stable even for an arbitrarily large initial disturbance. For $a > 0$ and $b < 0$ ('soft spring') the solution is only stable up to a critical value of the initial disturbance.

Dufour Effect *See* Soret Effect

Duhamel's Theorem
Let **L** be a linear operator whose coefficients and derivatives do not

88

involve the time variable t, and $\Phi(x, t)$ be the solution of the differential equation

$$\mathbf{L}\,\Phi + A(x)\,\frac{\partial^2 \Phi}{\partial t^2} + B(x)\,\frac{\partial \Phi}{\partial x} = 0 \qquad (0 < x < \mathbf{L}, t > 0)$$

having boundary conditions

$$\Phi(x, 0) = 0 \quad \frac{\partial \Phi}{\partial t} = 0 \qquad (0 < x < \mathbf{L}, t = 0)$$

$$a\frac{\partial \Phi}{\partial x} + \beta\Phi = b(t) \qquad (x = 0, t > 0),$$

there being as many homogeneous boundary conditions for $x = 0$ and/or $\mathbf{L}$ as are required for solution.

Then if $\chi(x, t)$ is the solution for the special case of $b(t) = 1$ $(t > 0)$, the function $\Phi(x, t)$ is given by

$$\Phi(x, t) = \chi(x, t)\,b(0) + \int_0^t \chi(x, t - \tau)\,b'(\tau)\,d\tau$$

$$= \frac{\partial}{\partial t} \int_0^t \chi(x, t - \tau)\,b(\tau)\,d\tau \qquad (0 < x < \mathbf{L}, t > 0)$$

Duhem-Margules Equation

This equation shows the connection between the relative amounts of two constituents of a system and their partial vapour pressures p_A and p_B. It is assumed that the vapours behave ideally but no assumption is made about the ideality of the liquids. The equation has the form

$$x_A\,\frac{d \ln p_A}{dx_A} = x_B\,\frac{d \ln p_B}{dx_B}$$

or

$$\frac{d \ln p_A}{d \ln x_A} = \frac{d \ln p_B}{d \ln x_A}$$

where x_A and x_B are the respective mole fractions of the two constituents.

89

Duhring Rule

According to U. Duhring if the boiling points of two substances A and B are T_A and T_B at pressure p, and T_A' and T_B' at pressure p'

$$\frac{T_A - T_A'}{T_B - T_B'} = \text{constant}$$

Dulong and Petit's Law

The specific heats of the elements are inversely proportional to their atomic weights. The atomic heats of solid elements are constant and approximately equal to 6·3 cal $^0K^{-1}$ mole^{-1}. Certain elements of low atomic weight and high melting point have, however, much lower atomic heats at ordinary temperatures. (*See* **Einstein's Equation for Specific Heat.**)

Dupré Equation

The work of adhesion W_{LS} at a gas-solid-liquid interface may be expressed in terms of the surface tensions of the three phases (γ_{LS}, γ_{GL} and γ_{GS}) by the equation

$$W_{LS} = \gamma_{GS} + \gamma_{GL} - \gamma_{LS}$$

E

Earnshaw's Theorem
The electric potential cannot be a maximum or a minimum at a point not occupied by charge. It follows that a small charged particle introduced into an electrostatic field cannot rest in stable equilibrium. The result follows from **Laplace's Equation**.

Edser and Butler's Bands
Dark bands, with constant frequency separation, in the spectrum of white light which has traversed a thin parallel-sided plate of a transparent material.

Eggertz's Method
A colorimetric estimation of carbon in steel, by dissolving the metal in nitric acid and comparing the colour to that produced by a similar metal of known carbon content.

Ehrenfest's Adiabatic Law
For a virtual, infinitely slow, alteration of the coupling conditions in a system, the quantum numbers do not change. The law is applicable both in the **Bohr Theory** and in quantum mechanics. The converse states that only such magnitudes that remain invariant for adiabatic changes can be quantized.

Ehrenfest's Equations
A second-order transition is characterized by discontinuous changes in the second-order derivative of the **Gibbs' Function**. The change of pressure with temperature during such a transition is given by

$$\frac{\mathrm{d}p}{\mathrm{d}T} = \frac{C_p^f - C_p^i}{TV\left(\gamma^f - \gamma^i\right)} = \frac{\gamma^f - \gamma^i}{\kappa^f - \kappa^i}$$

where f and i refer to the two phases. C_p is the specific heat at constant pressure, γ is the coefficient of volume expansion and κ is the compressibility.

Ehrenfest's Theorem
If the position and momentum of a classical particle are replaced by their quantum mechanical expectation values then the motion of a

91

quantum mechanical wave packet satisfying **Schrödinger's Equation**
will be equivalent to that defined by the classical equations of motion
for the particle, provided any potentials acting upon the particle do
not change over the dimensions of the wave-packet.

Ehrenhaft Effect
A helical type movement of fine particles in the lines of force of a
magnetic field during irradiation by light. The movement is similar to
that of electrically charged particles in an electric field and although the
motion was originally thought to be due to 'magnetic charges' on the
particles, it is caused by radiation pressure.

Einstein Coefficients
Suppose that in an atom two electronic states n and m exist ($m < n$)
such that a spontaneous transition can occur from n to m with the
emission of a photon. The probability that this transition will occur
was given by Einstein the symbol A_{nm}. In the presence of an external
radiation field of energy density $\sigma(\nu)$ where ν is the transition
frequency, stimulated transitions may occur from m to n.
Coefficients are defined for these stimulated transitions such that the total
probabilities of transition are given by $A_{nm} + B_{nm}\,\sigma(\nu)$ and $B_{mn}\,\sigma(\nu)$.
Einstein showed that the coefficients are related by the formulae

$$B_{mn} = \frac{g_n}{g_m} B_{nm} \quad \text{and} \quad A_{nm} = \frac{8\pi\nu^3 h}{c^3} B_{nm}$$

where g_n and g_m are the statistical weights of the two states n and m
respectively.

Einstein de Haas Effect *See* **Barnett Effect**

Einstein Mass-Energy Equation
A fundamental equation of modern physics connecting the mass m of a
particle with its energy E;

$$E = \tfrac{1}{2}\, m\, c^2.$$

Einstein Relation (for the Mobility of an Ion)
A relation connecting the diffusion coefficient D and mobility μ of an
ion of the form

$$\mu k T = e\, D.$$

Einstein's Equation for Specific Heat

It can be deduced from the classical atomic model that the specific heat at constant volume is connected to the gas constant R by the equation

$$C_V = 3R$$

This is not true, however, as all specific and hence atomic heats are temperature dependent. By application of quantum mechanical methods to this problem Einstein arrived at the equation

$$C_V = 3R\left(\frac{h\nu}{kT}\right)^2 \frac{\exp h\nu/kT}{(\exp h\nu/kT - 1)^2}$$

where ν is the Einstein frequency.

For most elements $h\nu/kT$ is sufficiently small for C_V to equal $3R$. Deviations from the Law of Dulong and Petit are due to a high value of ν. *See also* **Debye Equation.**

Einstein's Law

In the emission of electrons from metals by the incidence of radiation, if W is the work required to pass the electron through the surface, the maximum energy of the emitted electron is given by

$$\tfrac{1}{2}mv^2 = h\nu - W$$

where m is the mass and v the velocity of the electron, ν is the frequency of the radiation used and h is Planck's constant. W is known as the 'Work Function' of the metal.

Otherwise expressed

$$V_e = h\nu - h\nu_0$$

where V_e is the magnitude of the retarding potential required to decrease the velocity of the electron to zero and $\nu_0 = W/h$ is the minimum frequency which causes photoelectric emission from the given surface and is known as the 'Photoelectric Threshold'.

Einstein's Principle of Relativity

A theory of dynamics involving a space-time continuum and based on two major postulates:

I. Uniform motion of translation cannot be measured or detected by an observer stationed on a system of coordinates for measurements confined to that system.

II. The velocity of light in space is constant and independent of the relative velocities of the source and the observer.

These postulates apply to the Special Theory of Relativity where inertial frames move with constant velocities with respect to each other.

Einstein Summation Convention
In matrix and tensor notation whenever a letter occurs as a suffix twice in the same term, summation with respect to that suffix is to be automatically understood.

Elbs Reaction
A polynuclear hydrocarbon may be formed by the pyrolysis of a diaryl ketone containing a methyl group in an o-position to the carbonyl group.

Elster and Geitel Effect
In the presence of a gas, a heated conductor assumes a charge either positive or negative. In a vacuum the charge is always negative.

Encke Roots
Let the roots of

$$x^2 + a_1 x + a_2 = 0$$

be x_1 and x_2 where $|x_2| > |x_1|$. These roots, with their signs changed, are known as the Encke roots of the equation.

Eotvos Rule
This rule states that the rate of change of molar surface energy with temperature is the same for all liquids and independent of temperature. Ramsay and Shield found the coefficient of proportionality for certain organic liquids to be $2 \cdot 1$. This value of the constant is known as the Eotvos-Ramsay-Shield Constant. The constant varies for associated liquids and the variation was used to calculate the degree of association. Walden showed, however, that such calculations are unreliable as the coefficient varies greatly for different liquids and is, in general, not independent of temperature.

Eratosthenes, Sieve of

The prime numbers not greater than N can be found by writing all numbers up to N and then removing those numbers that are multiples of 2, 3, . . . etc., continuing until all the primes not greater than $\sqrt{N}$ have been removed.

Erichsen Value

An index of forming quality of sheet metal. The sheet is supported on a circular ring and deformed at the centre by a spherical pointed tool. The depth of impression in mm required to obtain fracture is the Erichsen value.

Erlenmeyer Azlactone Synthesis

The anhydrides of a-acylamino acids are formed by the condensation of an aromatic aldehyde with an aryl derivative of glycine in the presence of sodium acetate and acetic anhydride.

$$C_6H_5CHO + \underset{\underset{NH\,COR}{|}}{CH_2COOH} \xrightarrow[CH_3COO\,Na]{(CH_3CO)_2O} \begin{array}{c} C_6H_5CH{=}C-CO \\ \quad | \quad\;\; | \\ \quad N \quad O \\ \quad \;\backslash C / \\ \quad\;\; | \\ \quad\;\; R \end{array} + 2H_2O$$

where $R = CH_3$ or C_6H_5.

Eschweiler-Clarke Reaction

Formaldehyde and ammonium formate react to give triethylamine. In general, formaldehyde can be used to methylate primary and secondary amines in the presence of formic acid.

$$HCHO + HCOONH_3R \rightarrow CH_3NHR + CO_2 + H_2O$$

Essen Coefficient

A coefficient giving the torque per unit volume enclosed by the air gap of an electrical machine.

Ètard Reaction

Aromatic aldehydes may be prepared by oxidizing the alkyl benzenes with chromyl chloride in carbon tetrachloride solution. The complex which is precipitated is decomposed with water.

$$C_6H_5CH_3 + 2CrO_2Cl_2 \rightarrow C_6H_5CH_3.2CrO_2Cl_2 \xrightarrow{H_2O} C_6H_5CHO$$

Ettingshausen Effect

If a current is passed down a conductor which is placed in a magnetic field that is orthogonal to the direction of current flow, then a temperature gradient is established at right angles to both the current and the magnetic field, and is given by

$$\frac{dT}{dy} = P\,B_z\,I_x$$

where P is the Ettingshausen coefficient, B_z the magnetic flux density and I_x the current.

Euclidean Geometry

The study of ordinary two- or three-dimensional space with certain Euclidean axioms; e.g. the shortest distance between two points is a straight line, two halves of equal objects are equal, etc. It can also be used to mean the study of Euclidean space of any number of dimensions. This space may be defined as all sets of n numbers (x_1, x_2, ... x_n), where the distance $\rho\,(x, y)$ between $x = x\,(x_1, x_2 ... x_n)$ and $y = y\,(y_1, y_2, ... y_n)$ is defined as

$$\rho\,(x, y) = \left[\sum_{i=1}^{n} \left| x_1 - y_1 \right|^2 \right]^{1/2}$$

The distance is real or complex according as x is a real or complex number.

Euclid's Algorithm (for the Highest Common Factor)

If X and Y are polynomials of degree n and m respectively, where $n \geqslant m \geqslant 1$, then their highest common factor can be found as follows; Divide X by Y and obtain a remainder R_1, where R_1 is of degree $m_1 < m$. If $R_1 \neq 0$, divide Y by R_1 to obtain a remainder R_2 of degree $m_2 < m_1$ and divide R_1 by R_2. Continue until some $R_i = 0$ when R_{i-1} is the highest common factor or until R_i is a constant in which case the highest common factor is 1.

Eudoxus' Theorem *See* **Archimedes' Axiom**

Euler Diagram *See* **Venn Diagram**

Euler Force
The critical force required to cause the buckling of a beam supported at both ends. It is given by

$$\frac{\pi^2 E\, I}{l^2}$$

where E is **Young's Modulus** for the beam, I the moment of inertia about a horizontal axis through the centre of gravity and l the length. The beam may not necessarily completely collapse.

Eulerian Angles
In considering the kinematics of solid bodies, a set of angles serving both to define the orientation of the solid body and as independent variables in the equations of motion.

    If x, y and z are the three axes fixed in the body, and X, Y, Z represent a system of axes whose direction in space is fixed; θ is the angle between the Z and z axes. The line in which the xy plane cuts the XY plane is called the line of nodes and the angle between this and the x axis is termed ψ. The angle from the X axis measured in the positive direction round Z to the line of nodes is called ϕ. (Alternative definitions of the Eulerian angles are sometimes used.)

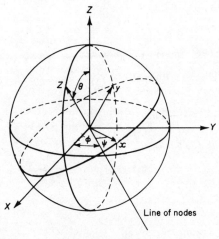

Eulerian Angles

97

Euler-Lagrange Equations
In variational calculus, the following problem is considered.

Given a function $Y(y, \frac{dy}{dx}, x)$ which is a function of y where y itself is a function of the independent variable x, it is required to minimize the variation of the following integral

$$\mathcal{L} = \int_a^b Y\left(y, \frac{dy}{dx}, x\right) dx$$

with respect to the choice of the function Y. The required Y which gives $\delta\mathcal{L} = 0$ is a solution of the Euler Lagrange differential equation

$$\frac{d}{dx}\left(\frac{\partial Y}{\partial \dot{y}}\right) = \frac{\partial Y}{\partial y}$$

where $\dot{y} = dy/dx$.

The **Lagrange Equations of Motion** for zero potential energy are a special case of the above.

Euler-Maclaurin Formula
If $f(x)$ is known to have the $(n + 1)$ values $y_0, y_1, \ldots, y_n$ at $(n + 1)$ points within the interval $(a, a + nh)$

$$\int_a^{a+nh} f(x)\, dx = h\left(\frac{y_0}{2} + y_1 + y_2 + \ldots + \frac{y_n}{2}\right)$$

$$- \sum_{\text{odd } r} \frac{h^{r+1}}{(r+1)!} B_{r+1}\left(y_n^{(r)} - y_0^{(r)}\right)$$

where $y_n^{(r)}$ and $y_0^{(r)}$ are the rth derivatives at the points a and $a + nh$, and B_{r+1} is the $(r + 1)$th Bernoullian number.

Euler's Constant (or Euler-Mascheroni Constant)
If $u_n = 1 + 1/2 + 1/3 + \ldots + 1/n - \ln n$, then, as $n \to \infty$, $u_n \to \gamma$, where γ is known as Euler's constant and is equal to $0 \cdot 577215 \ldots$

$$\text{Also } \gamma = - \int_0^\infty e^{-t} \ln t\, dt$$

$$= - \int_0^1 \ln \left\{\ln (1/s)\right\} ds$$

Euler's Definition of the Gamma Function

$$\Gamma(x) = \lim_{n \to \infty} \frac{n!}{x(x+1)(x+2)\dots(x+n-1)} \, n^{x-1}$$

Alternatively, the definition

$$\Gamma(x) = \int_0^\infty e^{-t} \, t^{x-1} \, dt$$

for $x > 0$ and real, is called Euler's Integral of the second kind.

Euler's Integral of the first kind is

$$\beta(p, q) = \int_0^1 t^{p-1} (1-t)^{q-1} \, dt$$

for $p > 0$, $q > 0$ and p and q real, where $\beta(p, q)$ is the Beta function given by

$$\beta(p, q) = \frac{\Gamma(p)\,\Gamma(q)}{\Gamma(p+q)}$$

Euler's Equation

In the motion of a perfect fluid

$$\frac{\partial v}{\partial t} + v \, \nabla v = F - \frac{1}{\rho} \, \nabla p$$

where v is the linear velocity of an element of the fluid at a point acted upon by a field of force F per unit mass of the fluid and also by the pressure p due to the remainder of the fluid. ρ is the density of the fluid.

Euler's Equations of Motion

For a rigid body fixed at one point and with coordinates fixed by the principal axes of the body, the equations of motion can be written as

$$I_{xx}\,\dot{\omega}_x + (I_{zz} - I_{yy})\,\omega_y\,\omega_z = L_x$$
$$I_{yy}\,\dot{\omega}_y + (I_{xx} - I_{zz})\,\omega_x\,\omega_z = L_y$$
$$I_{zz}\,\dot{\omega}_z + (I_{yy} - I_{xx})\,\omega_y\,\omega_x = L_z$$

where I_{ii} is the moment of inertia and L_j is the torque about a principal axis and ω_k is the angular momentum about the same reference line.

Euler's Formula (for a Polyhedron)
A formula connecting the number of faces (F), vertices (V) and edges (E) of any polyhedron whose faces do not intersect.

$$F + V = E + 2.$$

Euler's Kinematical Theorem
Every displacement of a rigid body is equivalent to a motion of translation plus a rotation about an axis passing through a point in the body.

Euler's Numbers
If the function $1/\cos Z$ is expanded into a Maclaurin's series

$$1 + \frac{E_1}{2!} Z^2 + \frac{E_2}{4!} Z^4 + \frac{E_3}{6!} Z^6 + \frac{E_n}{n!} Z^n + \ldots \ldots$$

E_n is called the nth Euler number.

Euler's Relation

$$e^{jkx} = \cos kx + j \sin kx$$

Euler's Theorem
If u is a homogeneous function of the nth degree of r variables $x_1, x_2, \ldots x_r$

$$(\text{ i.e. } u(ax_1, ax_2, \ldots ax_r) \equiv a^n u(x_1, x_2, \ldots x_r))$$

and is continuously differentiable, then

$$\left(x_1 \frac{\partial}{\partial x_1} + x_2 \frac{\partial}{\partial x_2} + \ldots + x_r \frac{\partial}{\partial x_r} \right)^m u = n^m u$$

where m may be any integer including zero.

Euler's Transformation (or Euler's Algorithm)

$$S = a_0 + a_1 x + a_2 x^2 + \ldots$$

$$= \frac{1}{1-x} a_0 + \frac{1}{1-x} \sum_{k=1}^{\infty} \left(\frac{x}{1-x} \right)^k \Delta^k a_0$$

where

$$\Delta^k a_0 = \sum_{m=0}^{k} (-1)^m \frac{k!}{m!(k-m)!} a_{k+n-m}$$

The second series may converge more rapidly than the first.

100

Everett's Interpolation Formula
A modified form of the **Newton-Gauss Interpolation Formula**.

Evershed Effect
Spectrographic evidence that the general motion of gases in the penumbral regions of sunspots is radially outwards. The radial velocities are greatest near the solar limb and least at the centre of the disc.

Ewald Sphere
The distance between planes (h, k, l) may be represented by d_{hkl} in a crystal lattice. For certain purposes it is convenient to construct the reciprocal lattice (*see for example* **Brillouin Zones**). Each point in this reciprocal lattice represents a family of planes in the *direct* lattice and it lies on the normal through the origin to the appropriate set of planes in the direct lattice and at a distance d_{hkl}^* inversely proportional to the distance between these planes. Thus we may write

$$d_{hkl}^* = \frac{K}{d_{(hkl)}}.$$

The constant K is quite arbitrary but, in practice, is usually made equal to unity or in x-ray diffraction to λ (the wavelength of the x-rays used). This second method has the advantage that the size of the 'mesh' of the lattice is independent of the wavelength used.

If we remember that $\sin_{(h00)} \theta = h \sin_{(100)} \theta$, $\sin_{(0k0)} \theta = k \sin_{(010)} \theta$ and $\sin_{(00l)} \theta = l \sin_{(001)} \theta$, it is easy to relate the lattice constants in the direct lattice a, b and c to the axial repeat distances in the reciprocal lattice a^*, b^*, c^*. The exact correspondence will, of course, depend upon K but enables a lattice to be constructed by measuring a given repeat distance along three axes at right angles to the crystal axes of the direct lattice.

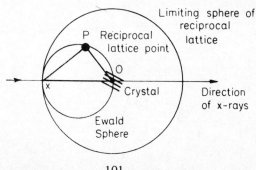

101

As each point in the reciprocal lattice represents a parallel set of planes in the real lattice, it is pertinent to ask which points in the reciprocal lattice will give a Bragg reflection for an x-ray beam of a given wavelength and direction with respect to the crystal axes in the direct lattice.

It is easy to show that if the direction of the x-ray beam is represented by OX in the figure and P is a lattice point, then the reciprocal lattice points which give a Bragg reflection lie on the circle through O and X with a radius equal to $1/\lambda$ for a value of K of unity. In three dimensions this circle becomes a sphere and is known as the **Ewald Sphere**. If the crystal is now rotated about some axis, then diffraction can occur provided the end of the scattering vector still lies on the surface of the corresponding Ewald sphere. Thus whatever the angle of the crystal, the reciprocal lattice is bounded for diffraction by another sphere called the limiting sphere.

These concepts simplify the interpretation of x-ray diffraction photographs and, indeed, of electron diffraction pictures also.

Eyring Equation
An equation in reaction kinetics connecting the rate of reaction R and the concentration of activated complexes C^*

$$R = C^* \left(kT/h \right) C_T$$

where the activated complexes are formed by collision of reacting molecules possessing more than a certain energy. kT/h is the vibration frequency of a bond in classical theory and C_T is a factor which accounts for the probability that the complex will break down into the products of the reaction rather than revert to the reactants.

Eyring Formula *See* **Norris-Eyring Reverberation Formula**

F

Faber Flaws
Regions of strain in superconductors that act as nucleation centres for
growth of superconducting regions.

Fabry-Perot Fringes
Fringes formed by interference between multiple beams reflected from
the inner silvered surfaces of two parallel plates. The fringes, formed at
equal inclination, consist of rings, and are similar in appearance to
Newton's Rings for monochromatic light. The higher the reflectivity of
the silvered surfaces, the sharper are the fringes. It is not practical to
observe white light fringes in this way.

Fahrenheit Scale
A scale of temperature with $32°$ at the freezing point and $212°$ at the
boiling point of water at normal pressure. $32°F = 0°C$, $212°F = 100°C$.

Fajans' Precipitation Rule
A radio-element will be co-precipitated with another substance if the
conditions of its precipitation are such that the element would form a
sparingly soluble compound if it were present in weighable quantities.
   There are many exceptions to this rule, but, as modified by Hahn,
the following is strictly true:
   However greatly diluted, an element will be carried down by a
crystalline precipitate, provided that it can be built into the crystal
lattice of the precipitate. (*See* **Paneth's Adsorption Rule**.)

Farad *See* **Appendix**

Faraday *See* **Appendix**

Faraday Dark Space
In a discharge tube, the dark space occurring between the positive
column and the negative glow.
   As the pressure is gradually reduced in a discharge tube filled with,

say, air, the discharge along the tube can be separated into the following regions, reading from anode to cathode:

(1) The pink glow immediately around the anode.
(2) The positive column of luminous gas extending from the anode to the beginning of
(3) The Faraday dark space.
(4) The negative glow, a pale violet discharge abruptly terminating at the
(5) **Hittorf** or **Crookes Dark Space**.
(6) A red or orange glow adjacent to the cathode.

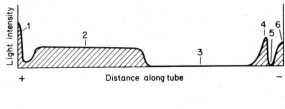

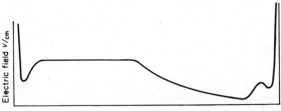

Faraday Dark Space

Faraday Effect

When a transparent isotropic medium is placed in a magnetic field it is capable of rotating the plane of polarization of the light travelling parallel to the lines of magnetic force. (*See also* **Verdet's Constant.**)

Faraday's Laws (of Electrolysis)

I. In electrolytic decomposition the number of ions charged or discharged at an electrode is proportional to the current passed.

II. The amounts of different substances deposited or dissolved by the same quantity of electricity are proportional to their equivalent weights.

104

Faraday's Laws (of Electromagnetic Induction)

I. When the flux of magnetic induction through a circuit is changing, an electromotive force is induced in the circuit.

II. The magnitude of the electromotive force is proportional to the rate of change of the flux.

Farey Sequence

All fractional (rational) numbers between 0 and 1 can be arranged in a sequence of groups such that if each number is denoted p/q then the numbers in each group have $(p + q)$ a constant $(= n$ where $n = 1, 2, \ldots)$. Thus the sequence is

$$\frac{0}{1}, \frac{1}{1}; \frac{1}{2}; \frac{1}{3}, \frac{1}{4}; \frac{2}{3}; \frac{1}{5}, \frac{1}{6}, \frac{2}{5}, \frac{3}{4}; \frac{1}{7}, \frac{3}{5}, \frac{1}{8}, \frac{2}{7}, \frac{4}{5}; \ldots$$

where each group is separated by semicolons. The group having $(p + q)$ $= n$ is called the Farey series of order n and each group is irreducible.

Favorskii Rearrangement

a-halogenated ketones in general undergo displacement of the halogen with great ease. Strong bases frequently yield acids or esters.

Fechner's Law

The eye can distinguish differences in illumination which are a constant (approx. 0.01) fraction of the total illumination.

Fehling's Solution

Aldehydes reduce this solution which contains a copper complex of tartaric acid in alkaline solution to red cuprous oxide.

Solution A: 70 gm $CuSO_4.5H_2O$ per litre.

Solution B: 120 gm NaOH per litre.
                350 gm $NaKC_4H_4O_6$ (Rochelle Salt) per litre.

Féjer's Theorem

A **Fourier Series** for the function $f(x)$ is summable for all points x at which two limits $f(x \pm 0)$ exist.

Fermat's Last Theorem

The equation $x^n + y^n = z^n$ where $x, y, z, \neq 0$ and $n > 2$ is impossible for integers x, y and z and n. This has never been proved for all values of n, although Fermat stated that he had a proof which was, however, not disclosed.

Fermat's Principle
When light passes from one point to another by reflection at a surface, the path taken is that which will be traversed in the least time. A similar law governs the refraction of light but in some cases leads to a maximum time instead of a minimum.

Fermat's Theorem
If p is a prime, and a is any number prime to p (i.e. not divisible by p) then

$$(a^{p-1}) - 1$$

is divisible by p.

Fermi *See* Appendix

Fermi Constant
An interaction or coupling constant for interactions involving a total of four Fermi-Dirac particles including the decay products. It has a value of approximately 10^{-49} erg cm^3. Examples of Fermi-Dirac particles that may be involved are the neutron, proton, electron, neutrino or muon.

Fermi-Dirac Statistics
The study of the probability of occupation of energy states in a quantized system by indistinguishable non-interacting particles which are subject to the **Pauli Exclusion Principle** so that there is a maximum of one particle per state. The probability that an energy state E is occupied is given by the **Fermi-Dirac Distribution Law**;

$$f(E) = \left\{ e^{(E - E_F)/kT} + 1 \right\}^{-1}$$

where E_F is called the **Fermi Energy** and is the energy at which the probability of occupation is ½. Fermi-Dirac statistics have particular application to electrons in metals. Particles which are subject to Fermi-Dirac statistics are called Fermi-Dirac particles or **fermions**. (Compare with **Bose-Einstein Statistics**.)

Fermi Level (Fermi Energy) *See* Fermi-Dirac Statistics

Fermi Plot *See* Kurie Plot

Fermi Selection Rules (for Beta Decay)

Fermi suggested that if the total angular momentum of a nucleus is I, then for the emission of a beta particle $\triangle I = 0$ (the emitted beta particle and neutrino have opposite spins) and there is no change of parity (i.e. no change of right-left symmetry). It was later found that the beta particle and the neutrino could have parallel spins giving rise to the **Gamow-Teller Selection Rules**;

$\triangle I = \pm 1$ or 0 (units of $h/2\pi$) with no change of parity, but with $I_i = 0$ to $I_f = 0$ not allowed where subscript i refers to the initial momentum and f the final momentum. Beta decay is a so-called 'weak interaction' and it is now known that parity is not always conserved in weak interactions.

Fermi's 'Golden' Rule

If a quantum system has two energy levels and is subject to an external perturbation, the transition probability between the lower state and the upper state at resonance is found to be proportional to the square of the matrix element of the perturbing term and to the density of final states available; i.e.

$$\rho_{kn} = \frac{\pi \, |H'_{kn}|^2}{\hbar} \rho \, (E_k)$$

where k and n are the final and initial states, H'_{kn} is a matrix element obtained by integration over spatial coordinates and $\rho \, (E_k)$ is the density of final states. For resonance, the applied frequency must equal $\omega_{kn} = (E_k - E_n)\hbar$ where E_k and E_n are the energies of states k and n respectively.

Ferranti Effect

The rise in voltage which takes place at the end of a long transmission line when the load is disconnected.

Feuerbach Circle (Nine-point Circle)

If AD, BE and CF are the altitudes of triangle ABC, H is the orthocentre and X, Y, Z, P, Q and R are the mid-points of BC, CA, AB, HA, HB, and HC respectively, then the circle through X, Y and Z also passes through P, Q, R, D, E and F.

Feynman Diagram

A method used in calculations involving interactions between many bodies (for instance nuclear particles) to show how a propagating particle starting at point r_1 at time t_1 moves to r_2 at time t_2 interacting during its passage with one or more other particles in the system. The propagation is divided into the probabilities for different types of interaction using the notation:

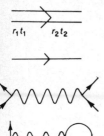

Total motion of a free particle from r_1 to r_2 in time t_1 to t_2.

Probability of free motion from r_1 to r_2.

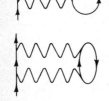

Probability of interaction between a particle at r_1 and a particle at r_2.

Probability of instantaneous forward scattering with no change of momentum.

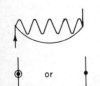

Probability of a chain of collision processes in which momentum at the beginning equals momentum at the end. (Analogous to the flow of current in an electrical network where 'flow in' equals 'flow out'.)

Probability of exchange scattering where the incoming particle is exchanged for the target particle.

Scattering at 0.

The important feature of a Feynman diagram is how the various lines are connected; i.e. the topology of the diagram. (For a detailed introduction, see R. D. Mattuck, *A Guide to Feynman Diagrams in the Many-Body Problem*, McGraw-Hill, 1967.)

Fick's Law
The quantity of material dv diffusing across a plane of unit area in time dt when the concentration gradient is $-(dc/dx)$ is given by

$$dv = -D\frac{dc}{dx}\,dt$$

where D, the diffusion coefficient, is the weight of material diffusing across unit area plane in unit time under unit concentration gradient.

Fieser's Solution
Oxygen may be removed from nitrogen by passing it through Fieser's solution. This solution is prepared by dissolving 20 gm of potassium hydroxide in 100 c.c. of water, adding 2 gm of sodium anthraquinone

β-sulphonate and 15 gm of 85% sodium hyposulphite (sodium thiosulphate.) The red solution is ready for use when it cools to room temperature and will absorb about 750 c.c. of oxygen.

Fischer-Hepp Rearrangement

The N-nitroso derivative of a secondary amine will rearrange to give the o- and p-nitroso amine

This reaction is a specific case of a more general reaction. The compound

where Z is an alkyl halogen, amino, nitroso or nitro-group will also undergo rearrangement in the same way.

Fischer-Indole Synthesis

This is the most important method of preparing indole derivatives. If the phenylhydrazine, or substituted phenylhydrazone of the appropriate aldehyde, ketone or ketonic acid is heated with an alkaline solution of picric acid, in the presence of zinc chloride as catalyst, ring closure is effected.

Fischer Polypeptide Synthesis

The first general method of synthesizing polypeptides in which an a-halogenated acid halide is allowed to react with an amino acid. The a halogen atom is subsequently replaced by an amino group. The dipeptide so formed is then reacted with another a-halogenated acid chloride to give a tripeptide. Fischer was able to produce a polypeptide containing eighteen amino acid residues. It is now only of historic interest as it cannot be used with amino acids which contain other functional groups.

Fischer-Speier Esterification

Aliphatic esters can be prepared by refluxing the respective alcohol and acid together. Such a reaction mixture may take several days to reach equilibrium. Fischer and Speier introduced the use of a mineral acid as a catalyst. Equimolar quantities of organic acid and alcohol give a poor yield and it is usual to have the acid in considerable excess. The yield is good with primary alcohols, fairly good with secondary alcohols and poor with tertiary alcohols.

Fischer-Tropsch Gasoline Synthesis

Water gas mixed with half its volume of hydrogen heated at $200°$–$300°C$ under a pressure of 1–200 atmospheres yields a mixture of hydrocarbons. The best catalyst is cobalt 100 parts, thoria 5 parts, magnesia 8 parts, kieselguhr 200 parts but the reaction gas must be freed from sulphur which poisons the catalyst.

Fisher's z Distribution

If s_1 and s_2 are two independent estimates of the variance (square of the standard deviation) of a normal or Gaussian distribution (see **Gauss' Error Curve**) then

$$z = \tfrac{1}{2}\ln (s_1/s_2)$$

and is a measure of the random sampling distribution.

Fittig Reaction

This reaction is analogous to the **Wurtz Reaction**. When an aryl halide is treated with sodium in ethereal solution the diaryl is formed.

$$2C_6H_5Br + 2Na \rightarrow C_6H_5.C_6H_5 + 2NaBr \ (20–30\%).$$

110

As a side reaction product o-diphenyl benzene and triphenylene are obtained.

Fitzgerald-Lorentz Contraction Hypothesis

In order to account for the null result of the Michelson-Morley experiment to determine the actual velocity of ether drift at any point, Fitzgerald in 1893 and Lorentz in 1895 suggested independently that the failure to detect velocity of motion through the ether was due to shrinkage of the apparatus used in the Michelson-Morley experiment by an amount which exactly concealed the motion.

Shortly, an arm of length l, when at rest, would, when moving with velocity u longitudinally through the ether, contract in a ratio $(1 - u^2/c^2)^{1/2}$ as a result of its motion.

Fizeau Fringes

Interference fringes observed in a wedge-shaped film due to interference between waves reflected from the two surfaces. With an extended white light source, the fringes coincide at the film and are said to be localized in the plane of the film. With a point source the fringes are no longer localized in this way.

Flade Potential

After a metal has become passivated, the decay of passivity with time is represented firstly by a steep fall of potential in the active direction, then by a less steep change and finally by a steep descent to the active value. The value of potential immediately preceding the final step is known as the 'Umschlagspunkt' or Flade potential.

Fleming's Rule

A rule relating the direction of flux, motion and e.m.f. in an electric machine. The forefinger, second finger and thumb placed at right angles to each other represent respectively the direction of flux, e.m.f. and torque. If the right hand is used, the conditions are as for a generator whereas the left hand gives the relation between these quantities for a motor.

Flood's Equation

In a binary fused salt system, say, $M_1A_1 + M_2A_2$, the liquidus temperature is given by

111

$$T_l = T_m \left(\frac{1 + X^2 \Delta G/L}{1 - [2RT_m \ln (1-x)]/L} \right)$$

where T_m is the melting point (degrees Absolute) of major component, M_1A_1,

ΔG is the standard free energy change of the reaction

$$M_1A_1 + M_2A_2 \rightleftharpoons M_1A_2 + M_2A_1$$

X is the mole fraction of the minor component, M_2A_2.

L is the molar heat of fusion of the major component, M_1A_1.

Floquet's Theorem *See* **Bloch's Functions.**

Flurscheim's Theory of Benzene Substitution

Flurscheim suggested in 1902 a method of determining the orientation of benzene substituents which was developed subsequently by other workers. If the mono-substituent in a benzene ring is a positive group then m-substitution takes place whereas if it is a negative group o-p substitution occurs.

Fokker-Planck Equation

A description of the time dependence of a **Markoff Process.** If $P(v, t/v_0)dv$ is the probability of a particle having a velocity between v and $v + dv$ at time t, when it is known to have a velocity v_0 at time $t = 0$, then

$$\frac{\partial P}{\partial t} + \frac{\partial}{\partial v}(M_1 P) - \tfrac{1}{2}\frac{\partial^2}{\partial v^2}(M_2 P) = 0$$

where M_1 and M_2 are given by

$$M_n = \frac{1}{\tau} < \left[\Delta v(\tau)\right]^n >$$

Also $< [\Delta v(\tau)]^n > = < [v(\tau) - v(0)]^n >$ is the nth moment of the velocity increment in time τ, and τ is a time interval which is infinitesimal when considering the macroscopic system, so that any terms involving M_n where $n > 2$ can be neglected. Note that in deriving the **Boltzmann Transport Equation** a particle can change its velocity abruptly as a result of a collision, whereas here a macroscopic particle can only change its velocity by a small amount in time τ.

Foreman's Method (for the Determination of Ammonia in Urine)
In the presence of a high concentration of alcohol and at an acidity corresponding to a faintly pink phenolphthalein the other nitrogenous bodies in urine, urea, etc., are not decomposed to ammonia whereas the ammonia from ammonium salts may be quickly and quantitatively removed by steam distillation.

Foster's Reactance Theorem
The most general driving point reactance function representing a network, in which every mesh contains independent inductance and capacitance, has the form

$$Z = jH \frac{(\omega^2 - \omega_1^2)(\omega^2 - \omega_3^2) \ldots (\omega^2 - \omega_{2n-1}^2)}{\omega(\omega^2 - \omega_2^2)(\omega^2 - \omega_4^2) \ldots (\omega^2 - \omega_{2n-2}^2)}$$

where H is a constant and the poles and zeros of the function are simple and alternate.

$$0 < \omega_1 < \omega_2 < \ldots < \omega_{2n-2} < \omega_{2n-1} < \infty$$

Foucault Current
A current induced in the interior of conductors by variations of magnetic flux.

Foucault's Pendulum
A freely suspended pendulum does not continue to swing in the same direction. The plane of oscillation appears to rotate through about $12°$ per hour in our latitudes in an opposite sense to the rotation of the earth. In fact, the plane of oscillation remains constant and it is the revolution of the earth about its axis which gives rise to the apparent rotation. For other latitudes, ψ the angle of rotation per hour is given by

$$\psi = 15 \sin \phi$$

where ϕ is the geographical latitude of the observer.

113

Foulger's Test for Fructose

Foulger's reagent is prepared as follows: 40 gm of urea is added to 80 c.c. of 4% v./v. sulphuric acid and then 2 gm stannous chloride are added. The solution is boiled until clear and made up to 100 c.c. with the sulphuric acid. In the presence of fructose, a blue coloration is developed.

Fourier Bessel Integral

An expression corresponding to the Fourier Integral, but with exponential kernels replaced by Bessel Functions.

$$f(x) = \int_0^\infty J_m(kn)k\,dk \int_0^\infty f(\xi)J_m(k\xi)\,\xi\,d\xi$$

Fourier Integral

The Fourier series

$$f(x) = \sum_{n=0}^\infty \epsilon_n \left\{ \int_{-\pi}^{\pi} f(z) \cos mz\,dz \right\} \cos mx$$

$$+ \sum_{n=1}^\infty \epsilon_n \left\{ \int_{-\pi}^{\pi} f(z) \sin mz\,dz \right\} \sin mx$$

where

$$\begin{aligned}\epsilon_n &= 1/2\pi \text{ when } n = 0 \\ &= 1/\pi \text{ when } n \neq 0\end{aligned}$$

which holds for $-\pi < x < \pi$, becomes, in the limit, for $-\infty < x < \infty$, and for continuous $m \to k$, the Fourier Integral

$$f(x) = \frac{1}{2\pi} \int_{-\infty}^{\infty} e^{jkx}\,dk \int_{-\infty}^{\infty} f(\xi)\,e^{-jk\xi}\,d\xi$$

Fourier Number

A dimensionless number used in heat transmission and given by $\lambda t / C\rho l^2$ where λ is thermal conductivity, C specific heat, ρ density and l a characteristic length.

Fourier's Law (of heat conduction)
Heat flux in an isotropic medium is given by $-\lambda(dT/dx)$ where λ is the thermal conductivity and dT/dx is the temperature gradient.

Fourier's Series
Any function of x, say $f(x)$, can be expressed as the sum of a series of sines and cosines.

$$\tfrac{1}{2}b_0 + b_1 \cos x + b_2 \cos 2x + \ldots$$
$$+ a_1 \sin x + a_2 \sin 2x + \ldots$$

where

$$b_m = \frac{1}{\pi} \int_{-\pi}^{+\pi} f(x) \cos mx \, dx$$

$$a_m = \frac{1}{\pi} \int_{-\pi}^{+\pi} f(x) \sin mx \, dx$$

and this development holds for all values of x between $-\pi$ and $+\pi$ providing $f(x)$ is monotonic and is finite and continuous in the interval $-\pi < x < +\pi$, or, if discontinuous, has only finite discontinuities.

Fourier's Series was developed in the process of finding a solution to an equation of the type

$$\frac{\partial^2 u}{\partial x^2} + \frac{\partial^2 u}{\partial y^2} = 0$$

which arose in a problem connected with the distribution of heat in a solid conducting medium.

Fourier Transform
If $f(x)$ is such that

$$\int_{-\infty}^{\infty} |f(x)|^2 \, dx$$

is finite, and if

$$F(k) = \frac{1}{\sqrt{(2\pi)}} \int_{-\infty}^{\infty} f(x) \, e^{jkx} \, dx$$

then $F(k)$ is called the Fourier transform of $f(x)$ and

$$f(x) = \frac{1}{\sqrt{(2\pi)}} \int_{-\infty}^{\infty} F(k) \, e^{-jkx} \, dk$$

Franck-Condon Principle

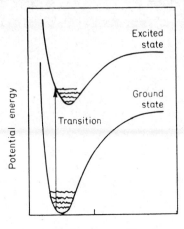

Franck-Condon Principle

Electronic transitions in molecules occur in times which are very short in comparison with the period of vibration of the nuclei. Hence in an energy level diagram the most probable electronic transitions occur vertically.

Frankland's Method
Dialkyl zinc compounds readily react with alkyl halides to form hydrocarbons. The dialkyl zinc compounds are difficult to handle and this method is, in general, only used to form paraffins containing a quaternary carbon atom.

$$(CH_3)_3C.Cl + (CH_3)_2Zn \rightarrow (CH_3)_4C + CH_3ZnCl$$

Compare **Wurtz Reaction**.

Franklin Equation
Relates the energy level of sound in a room as a function of time after the source has been cut off.

$$E = E_0 \exp(-caS/4V \times t)$$

where E and E_0 are the final and initial sound levels, c the velocity of sound, S the surface area, V the volume of the room, t the time and a is the mean sound absorption coefficient.

116

Frank Partial Dislocation *See* **Shockley Partial Dislocation**

Frank-Read Source

A source of dislocation loops in a strained crystal. If a dislocation is pinned at two points (this can arise from two successive cross-slip processes), then the dislocation can bow out under stress through stages 2 to 4 until it produces the complete loop 5; whereupon the production of a new loop can begin.

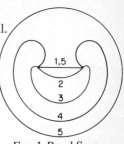

Frank-Read Source

Franz-Keldysh Effect
A shift to longer wavelengths in the absorption edge of semiconductors produced by a strong electric field. Thus, optical transmission and reflectance can be modulated by the application of an electric field.

Frasch Sulphur Process
In Louisiana and Texas the sulphur beds occur beneath 400 ft of clay and quicksands and 90 ft of limestone. Mining of this very rich deposit was not possible by ordinary methods because of the large quantities of water in the subsoil. The difficulty was solved in a most ingenious way by Frasch. He devised a process whereby superheated water was forced under pressure into the sulphur bed and the molten sulphur was forced to the surface.

Fraunhofer Diffraction
Diffraction phenomena may be divided into two classes:
  I. When the light source and the screen upon which observations of the pattern are taken are at an infinite distance from the aperture, the diffraction is known as Fraunhofer diffraction.
  II. When either the source or the screen or both are at a finite distance then it is known as **Fresnel Diffraction.**

Fredholm Equation
An integral equation of the same type as the **Volterra Equation** with constant limits of integration.

Frenkel Defect

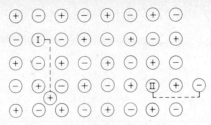

Frenkel Defect

There exists in a crystal in thermal equilibrium a number of vacant lattice points. An ion removed from a lattice position and placed in an interstitial position in the lattice leaves behind a Frenkel defect. If, however, the displaced ion is removed to the surface of the crystal the defect is known as a **Schottky Defect**.

Frenkel Exciton

An exciton is an excitation which can occur in a crystalline solid and which may be considered as a conduction-band electron and a valence-band hole bound together (although with finite separation) moving through the crystal. A Frenkel exciton is a tightly bound electron-hole pair with small separation of the electron and hole. A **Wannier** (or **Mott**) **Exciton** is a weakly bound pair with large separation.

Fresnel-Arago Laws (of Polarized Interference)

I. Two rays polarized at right angles do not interfere.

II. Two rays polarized at right angles obtained from the same beam of polarized light will interfere in the same manner as ordinary light only when brought into the same plane.

III. Two rays, polarized at right angles and obtained from perpendicularly polarized components of unpolarized light, never interfere even after rotation of their planes of polarization.

Fresnel Diffraction *See* **Fraunhofer Diffraction**

Fresnel Integrals

The evaluation of the intensity of a diffraction pattern leads to an expression containing two integrals of the form:

$$\int \cos \frac{\pi v}{2} \, dv \text{ and } \int \sin \frac{\pi v}{2} \, dv$$

where v depends upon the distance of the aperture from the source and from the screen and upon the displacement of the ray. These two integrals represent the components, along two rectangular axes, of the resultant amplitude and may be evaluated between given limits by the use of tables calculated by Fresnel and other workers.

Fresnel Reflection Formulae

For incident light of intensity I_0 polarized in the plane of incidence (i.e. with the magnetic vector in the plane of incidence), the intensity of light reflected from the surface of a transparent medium is

$$I = I_0 \frac{\sin^2 (i - r)}{\sin^2 (i + r)}$$

where i and r are the angles of incidence and refraction respectively. For light polarized perpendicular to the plane of incidence, the intensity of the reflected light is

$$I' = I_0 \frac{\tan^2 (i - r)}{\tan^2 (i + r)}.$$

For unpolarized light of incident intensity I_0, the reflected intensity is $\frac{1}{2}(I + I')$. For light polarized or unpolarized at normal incidence, passing from a medium of refractive index n to one of refractive index n', the intensity of the reflected light is

$$I_0 \left(\frac{n' - n}{n' + n} \right)^2.$$

Fresnel Zone

In considering the effect of a wave front at any point P it is convenient to separate the wave front into regions formed first by constructing a tangent sphere to the wave front with centre at P and radius, say, a. Successive curves are then described on the wave front by spheres of radius $a + \frac{1}{2}\lambda$, $a + 2.\frac{1}{2}\lambda$, $a + 3.\frac{1}{2}\lambda$, etc., where λ is the wavelength and the zones enclosed between any two adjacent curves are called Fresnel Zones. It can easily be shown that the areas of these zones are equal.

Freundlich Isotherm

According to Freundlich the variation in adsorption with pressure at constant temperature may be represented by the equation

$$\frac{x}{m} = kp^{1/n}$$

119

where x is the mass of gas adsorbed by m grammes of adsorbent at a pressure p; k and n being constants for the system.

Freund Method (for Preparation of *Cyclo*paraffins)
When $a\omega$ dihalogen derivatives of the paraffins are treated with sodium or zinc, *cyclo*paraffins are formed

$$\begin{array}{c} CH_2Br \\ / \\ CH_2 \\ \backslash \\ CH_2Br \end{array} + Zn \rightarrow \begin{array}{c} CH_2 \\ / \\ CH_2 \\ \backslash \\ CH_2 \end{array} + ZnBr_2$$

Compare **Wurtz Reaction**.

Friedel-Crafts Reaction
This reaction is one of the most famous and useful reactions in organic chemistry. It consists of the condensation of an alkyl chloride, or an alkyl or aryl acid chloride with an aromatic hydrocarbon in the presence of aluminium chloride. Thus a simple example of the hydrocarbon reaction is

$$\bigcirc + CH_3Cl \xrightarrow{AlCl_3} \bigcirc^{CH_3} + HCl$$

A disadvantage of this reaction is, however, that polysubstitution can also occur and hence in the presence of excess methyl chloride benzene will be converted not only into toluene and xylene but to the higher methyl benzenes, e.g. durene. Other anhydrous inorganic halides have been used as the catalyst but they are less efficient. The relative potencies of these catalysts are, for example,

$$AlCl_3 > FeCl_3 > SnCl_4 > BF_3 > ZnCl_2$$

The ketone synthesis is free from the complicated side reaction of the hydrocarbon synthesis and in this case the reaction between the acid chloride and the hydrocarbon occurs in the presence of a full equivalent of aluminium chloride. The reaction is usually exothermic to start with and the reactants are cooled, but the condensation is completed by refluxing the reactants together.

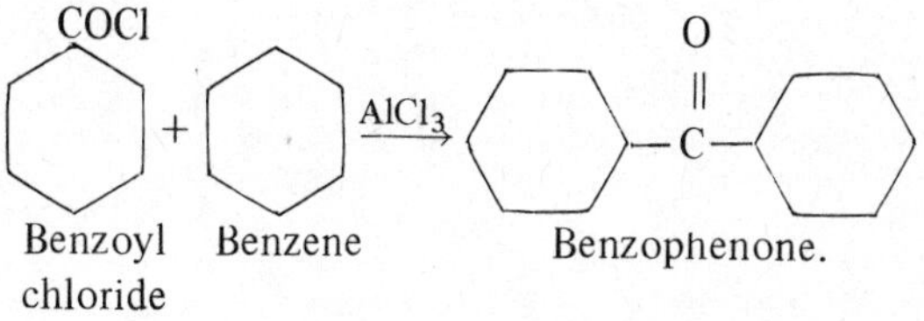

Benzoyl chloride   Benzene   Benzophenone.

The ketones produced by this reaction may be reduced to the hydrocarbon by means of the **Clemmensen Reduction.**

Friedel's Law
Every crystal diffracts x-rays as if a centre of symmetry were present. Thus the diffraction symmetry of a crystal is its point group symmetry plus a centre of symmetry, and corresponds to any one of eleven kinds of centrosymmetry (the **Laue symmetry groups**); i.e. the intensity of reflection from opposite sides of the same set of crystal planes is the same. The law can fail in polar crystals for radiation which is close in wavelength to an absorption edge.

Friedländer Synthesis
This synthesis is an important method of preparing quinoline and substituted quinolines. Thus when *o*-aminobenzaldehyde is condensed with acetaldehyde in the presence of aqueous sodium hydroxide, quinoline is formed:

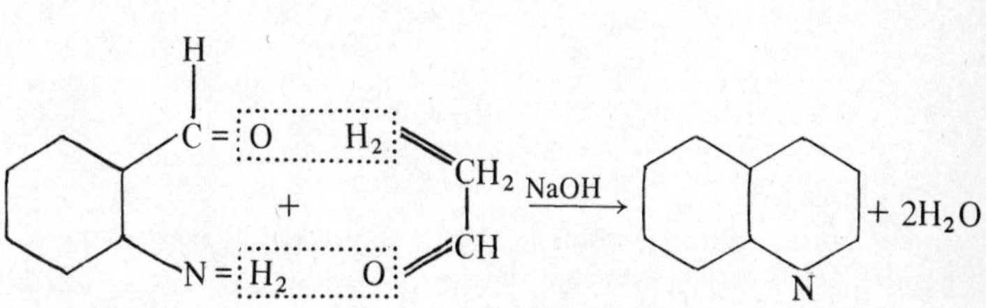

Fries Rearrangement
Phenyl esters when treated with anhydrous aluminium chloride rearrange to give a mixture of *ortho-* and *para-*hydroxyketones.

121

Generally low temperatures (60°C) favour the formation of the
p-isomer, whereas high temperatures (above 160°C) favour the
o-isomer.

Fries' Rule
The most stable arrangement of the bonds of a polynuclear compound
is that form which has the maximum number of rings in the benzenoid
form, i.e. three double bonds in each ring.

Frobenius' Method
A method of solving homogeneous linear differential equations of any
order, with variable coefficients, where the solution is assumed in the
form of a series

$$y = x^c(a_0 + a_1x + \ldots + a_nx^n + \ldots)$$

where $c, a_0, a_1, \ldots, a_n, \ldots$ are determined by substituting the series into
the equation and setting the complete coefficient of any power of x to
zero.

Froude's Curve
A curve whose offset is given by

$$y = \frac{x^3}{6lr}$$

where x is the distance from the tangent point, l is the length of
transition and r is the radius of the circular arc.

Froude's Number
A dimensionless number similar to the **Reynolds Number** for bodies
falling through a fluid and given by

$$F = \frac{v^2}{la}$$

where v is the velocity, l a characteristic length and a the acceleration
for each body.

G

Gabriel's Phthalimide Synthesis
Phthalimide is converted by means of alcoholic potassium hydroxide into the potassium salt. This salt, on heating with an alkyl halide, gives the N-alkyl phthalimide which may be hydrolysed by heating with 20% hydrochloric acid under pressure or refluxing with potassium hydroxide solution to give phthallic acid and the N-alkyl amine.

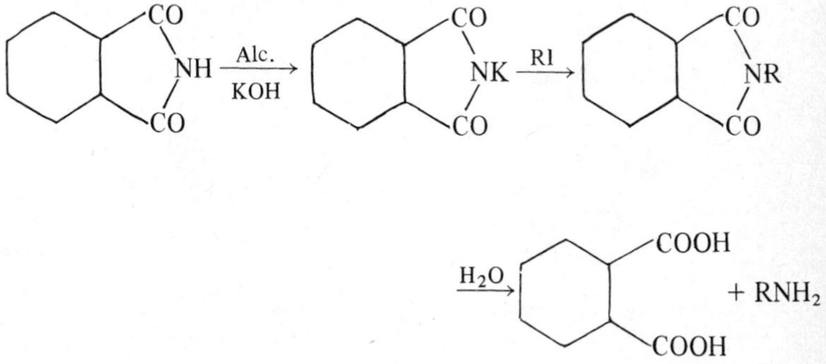

Galilean Transformation
A transformation of cartesian coordinates x, y, z, t given by the equations

$$x' = x - vt$$
$$y' = y$$
$$z' = z$$
$$t' = t$$

where v is the velocity (in the x direction) with which one system of coordinates moves with respect to the other.

Galileo *See* Appendix

Galtonian Curve
A curve showing the deviation of any characteristic from the normal.

Galvano- Effects

Named after L. A. Galvani, these are effects occurring when an electric current is passed through a conductor (*see the* **Thomson** and **Peltier Effects**). Similarly, galvanomagnetic effects occur when a magnetic field is also present (*see the* **Hall** and **Ettingshausen Effects**).

Gamow-Teller Selection Rules (for Beta Decay) *See* Fermi Selection Rules

Gatterman Aldehyde Synthesis

When benzene is treated with a mixture of hydrocyanic acid and hydrochloric acid in the presence of aluminium chloride a complex is formed which hydrolyses to benzaldehyde. The yield is poor. This synthesis is also applicable to phenols and phenolic ethers.

Gatterman Carbon Monoxide Synthesis of Aldehydes (Gatterman-Koch Reaction)

Formyl chloride $ClCHO$ is very stable at $77°K$. A mixture of carbon monoxide and hydrogen chloride in the presence of anhydrous aluminium chloride and cuprous chloride behaves like formyl chloride in reactions with aromatic compounds.

Gattermann Reaction

Chlorine or bromine may be introduced into the benzene or substituted benzene ring by diazotising the appropriate amine in the presence of the acid and decomposing the diazonium salt with copper powder.

Gauss *See* **Appendix**

Gauss' Analogies *See* **Delambre's Analogies**

Gauss' Error Curve
In a set of observations, where measurements are subject to random errors, the number of measurements associated with any given error may be expressed by a curve

$$y = \frac{h}{\sqrt{\pi}} e^{-h^2 x^2}$$

where y is the number of observations with deviation x from the mean value and h is a constant giving the measure of precision of the set of observations. Such a set, if it follows closely the distribution given by the above relation, is termed a normal distribution.

Gauss' Hypergeometric Differential Equation

$$(x^2 - x)\frac{d^2 y}{dx^2} + \left[(1 + a + \beta) x - \gamma \right] \frac{dy}{dx} + a\beta y = 0$$

with a, β and γ constants and γ not an integer. Series expansion about $x = 0$ gives a particular solution;

$$F(a, \beta; \gamma; x) = 1 + \frac{a\beta}{\gamma} x + \frac{1}{2!} \frac{a(a+1)\beta(\beta+1)}{\gamma(\gamma+1)} x^2 + \ldots$$

which converges uniformly for $|x| < 1$ and is called the hypergeometric series.

Gaussian Complex Integers
Complex numbers of the form $a + jb$, where a, b are rational integers and $j = \sqrt{-1}$.

125

Gaussian Curvature

At a general point on a surface there is a direction for which the radius of curvature of a normal section is a maximum and a perpendicular direction for which it is a minimum. The corresponding radii of curvature, ρ_1 and ρ_2 are called principal radii of curvature, and the total, or Gaussian, curvature is given by

$$\kappa = \frac{1}{\rho_1 \rho_2}$$

Gaussian Positions

Any point on the axis produced of a bar magnet can be said to be in Gauss A position and any point on the magnetic equator can be said to be in Gauss B position.

Gaussian Units

A composite system of units in which all electric quantities are measured in absolute c.g.s. electrostatic units and all magnetic quantities are measured in c.g.s. electromagnetic units.

Gauss' Multiplication Theorem

$$\Gamma\left(nz\right) = \sqrt{\left[\frac{n^{2nx-1}}{(2\pi)^{n-1}}\right]} \Gamma\left(z\right) \Gamma\left(z + \frac{1}{n}\right) \Gamma\left(z + \frac{2}{n}\right) \ldots \Gamma\left(z + \frac{n-1}{n}\right)$$

$$(n = 2, 3, \ldots)$$

See also **Euler's Definition of the Gamma Function**.

Gauss' Optics Formulae

I. For a single spherical surface; If s is the object distance, s' the image distance and n and n' the refractive indices of the two media separated by the spherical surface of radius of curvature r, then

$$\frac{n}{s} + \frac{n'}{s'} = \frac{n' - n}{r}$$

II. For a lens;

$$\frac{1}{s} + \frac{1}{s'} = \frac{1}{f}$$

where f is the focal length of the lens and there are the same media on either side of the lens. (f is positive for a converging system, negative for a diverging one; if s is positive when measured to the left, s' is positive when measured to the right.)

Gauss' Π Function

$$\Pi(k, z) = k^z \prod_{n=1}^{k} \frac{n}{z + n} \qquad \Pi(z) = \lim_{k=\infty} \Pi(k, z) = \Gamma(z + 1)$$

Gauss' Reciprocal Theorem

Let charges $e_1, e_2, \ldots$ be placed at $P_1, P_2, \ldots$ and let V_1 be the potential at P_1 due to all charges except e_1. Let $e_1', e_2', \ldots$ be any other charges placed at the same points and let $V_1', V_2', \ldots$ be the corresponding potentials.

Then

$$\Sigma eV' = \Sigma e'V$$

Gauss' Test (for Convergence or Divergence)

If for the series Σa_n the ratio of consecutive terms is given by

$$\frac{a_{n+1}}{a_n} = 1 - \frac{A}{n} + O\left(\frac{1}{n^{1+\lambda}}\right)$$

where $\lambda > 0$, then for $A > 1$, the series converges; if $A \leqslant 1$, the series diverges.

Gauss' Theorems

The surface integral of a vector over a closed surface is equal to the volume integral of the divergence of the vector throughout the enclosed volume.

$$\iiint \text{div } \mathbf{A} \, dV = \iint \mathbf{A} \, d\mathbf{S}$$

The magnetic flux $\mathbf{H}$ outwards through a closed surface $\mathbf{S}$ is 4π times the sum of the strength of all the magnetic poles, m, inside the surface.

$$\iint \mathbf{H} \, d\mathbf{S} = 4\pi m$$

127

Gay Lussac Law of Combining Volumes
If gases interact and form a gaseous product, the volumes of the
reacting gases and the volumes of the products are in simple proportion.

Gay Lussac's Law (of Expansion)
At constant pressure, the coefficient of thermal expansion (volume
coefficient) is the same for all gases and has a mean value $a =$
$0.0036604 = 1/273.2$.

Gegenbauer Polynomials
When n in the equation

$$(x^2 - 1) \frac{d^2 y}{dx^2} + 2(\beta + 1) x \frac{dy}{dx} - n(n + 2\beta + 1)y = 0$$

is zero or positive integer, the solution becomes a polynomial
(Gegenbauer polynomial); otherwise it is an infinite series. For $\beta = 0$,
the differential equation becomes Legendre's equation satisfied by
Legendre's Coefficients, and for $\beta = -\frac{1}{2}$ the equation is satisfied by
Chebyshev Polynomials. The Gegenbauer polynomials can be generated
as coefficients of the power series

$$\frac{2^\beta}{(1 - 2tx + t^2)^{\beta + \frac{1}{2}}} = \frac{\sqrt{\pi}}{\Gamma(\beta + \frac{1}{2})} \sum_{n=0}^{\infty} t^n C_n^\beta (x)$$

where $\Gamma(\beta + \frac{1}{2})$ is the gamma function for $\beta + \frac{1}{2}$.

Geiger-Nuttal Rule
H. Geiger and J. M. Nuttal discovered experimentally a relationship
between the radioactive constant λ and the range, R, of a-particles
emitted by a radioactive element.

$$\log \lambda = b + c \log R$$

where b and c are constants, b differs for each of the three radioactive
series and c is practically the same for each series.

Geigy-Hardisty Process
A process for the production of sebacic acid from castor oil or its acids
by treatment with caustic alkali at a high temperature.

Geitel Effect *See* **Elster and Geitel Effect**

Gerhardt's Test (for Acetoacetic Acid in Urine)

A. To 5 c.c. of urine in a test tube add ferric chloride solution dropwise until no more ferric phosphate is formed. Filter and add more $FeCl_3$. A Bordeaux-red coloration indicates acetoacetic acid.

B. If the first test is positive, shake 50 c.c. of urine plus 3 drops of conc. sulphuric acid with ether. Pipette off the ether and treat with very dilute ferric chloride. The lower layer becomes violet; add more ferric chloride, the colour changes to red.

Giauque's Temperature Scale

A temperature scale using only one fixed point, the triple point of water, which is defined as 273·16 degrees absolute. The scale was internationally accepted in 1954.

Gibbs' Adsorption Equation

J. W. Gibbs deduced thermodynamically the relationship between surface tension and adsorption. If Γ is the surface concentration of solute per unit area of interface, a the activity of the solute and γ the surface tension, then

$$\Gamma = -\frac{1}{RT}\frac{d\gamma}{d(\ln a)} = -\frac{a}{RT}\frac{d\gamma}{da}$$

Gibbs-Duhem Equation

A relation between the chemical potential and concentration of species in a mixture at constant temperature and pressure. If there are n_i moles of species i present having chemical potential μ_i, then

$$\sum_i n_i\, d\mu_i = \sum_i x_i\, d\mu_i = 0$$

where $x_i = \dfrac{n_i}{\sum_i n_i}$ is the mole fraction.

Gibbs' Function (or Free Energy)

The Gibbs' Function for an infinitesimal reversible process is given by

$$G = H - TS$$

where G is the Gibbs' Function or free energy, H is the enthalpy and S is the entropy.

129

Gibbs-Helmholtz Equation

This equation was deduced independently by J. W. Gibbs in 1875 and H. von Helmholtz in 1882. It may be expressed in a number of ways. If ΔG is the change in free energy during a reaction and ΔH the increase in heat content, then,

$$\Delta G - \Delta H = T \left[\frac{\partial \Delta G}{\partial T} \right]_P$$

Gibbs' Paradox

Experience suggests that as two diffusing gases become more and more alike the change in entropy due to diffusion should get smaller and smaller, approaching zero as the gases become identical. The fact that this is not the case is known as the Gibbs' Paradox. Bridgman explained it in the following fashion: in principle at least it is possible to distinguish the dissimilar gases by a series of experimental operations. In the limit, however, when the gases become identical there is a discontinuity in the operations in as much that no instrumental operation exists by which the gases may be distinguished. Hence a discontinuity in a function such as change in entropy is to be expected.

Gibbs' Phase Rule

$$F = C + 2 - P$$

where F is the number of degrees of freedom of a system (temperature, pressure and concentration) which must be fixed in order to define the system uniquely, C is the number of components, that is the smallest number of distinct substances which enable the constitution of each phase to be expressed and P is the number of phases of the system, i.e. the homogeneous, mechanically separable, physically distinct portions of the heterogeneous system.

Gibbs' Phenomenon

In considering the transient response of an electrical network to a unit step function the wider the band of frequencies passed by the network the better will the response be. Gibbs, however, stated that no matter how high the cut-off frequency of the network was, the maximum amplitude of the trailing overshoot of the leading edge never becomes less than 0·175. More generally, a **Fourier Series** expansion of a function round a discontinuity results in an overshoot even if an infinite number of terms is taken.

Gibbs' Rule
J. W. Gibbs proposed a more rigid form of **Dalton's Law of Partial Pressures**. The pressure of a mixture of gases is only equal to the sum of the partial pressures if the chemical potential of the gas is unchanged by being incorporated in a mixture.

Gilbert *See* Appendix

Giorgi System
A system of measurement in which the units are the metre, the kilogramme, the second and the ampere.

Gladstone-Dale Law
The refractive index of a substance varies with change in temperature or in volume according to the formula

$$k = \frac{n-1}{\rho}$$

where n is the refractive index, ρ the density and k is a constant.

Gmelin's Test (for Bile Pigments)
Take a few c.c. of fuming nitric acid and pipette carefully on to the surface an equal amount of bile. Shake the tube gently from side to side and note the sequence of colours as the bile becomes oxidized by the acid. Proceeding from the acid to the bile they are yellow, red, violet, blue and green.

   This test has been modified in various ways—e.g. the nitric acid is spotted on to a thin film of bile on a tile when concentric rings of these colours are formed.

Goldbach Conjecture
Any even number other than 2 can be represented as the sum of two prime numbers. It is unproved.

Goldschmidt's Law
After consideration of the crystal structures of a large number of

131

inorganic compounds V. Goldschmidt postulated in 1929 the fundamental law of crystal chemistry:

The structure of a crystal is determined by the ratio of the numbers, the ratio of sizes and the properties of polarization of its structural units.

Gomberg Hey Reaction

Unsymmetrical diaryls can be prepared by treating an aryl diazonium salt solution with sodium hydroxide or sodium acetate in the presence of a liquid aromatic compound.

Gomberg Reaction

When a solution of a diazonium salt is mixed with a liquid aromatic compound at low temperatures (5°C) and made alkaline with NaOH, a small yield (10–40%) of a product containing two united aromatic nuclei is obtained.

$$ArN_2X + HC_6H_4Y \xrightarrow{NaOH} ArC_6H_4Y + N_2 + NaX + H_2O$$

Gordon's Formula

An empirical formula giving the collapsing load P for a given column

$$P = \frac{\sigma A}{1 + c(l/d)^2}$$

where σ = sate compressive stress for very short lengths of the material,

A = cross-sectional area,
l = length of column,
d = least diameter of cross-section,
c = constant.

Graeffe's Method (for Determining Roots of an Equation)

Let

$$f(x) = a_0x^n - a_1x^{n-1} + \ldots + a_n = 0$$

be the equation. Form the equation $f(x) \cdot f(-x) = 0$ which is an equation of the nth degree in x^2. Continue this process to obtain an equation

$$P_0 z^n - P_1 z^{n-1} + P_2 z^{n-2} - \ldots + P_n = 0$$

whose roots are the 2^rth powers of the roots of the given equation. Put $\lambda = 2^r$. Let the roots of the given equation be

$$x_1 > x_2 > x_3 > \ldots > x_n$$

Then, for large values of λ,

$$x_1{}^\lambda = \frac{P_1}{P_0}, x_2{}^\lambda = \frac{P_2}{P_0}, \ldots x_n{}^\lambda = \frac{P_n}{P_{n-1}}$$

If the roots are real $x_1 = \lambda\sqrt{(P_1/P_0)}$, etc., and their sign is determined by trial in $f(x) = 0$. For treatment when the roots are complex, *see Encyklopadie der Math. Wiss.* I, **1**, 3a (Runge).

Graetz Number
A dimensionless number used in problems of heat transfer and defined as

$$\frac{v \rho A \, C_p}{\lambda \, l}$$

where v is the velocity of flow, ρ the density, A the surface area, C_p the specific heat at constant pressure, λ the thermal conductivity and l a characteristic length.

Graham's Law of Diffusion
At constant temperature and pressure, the rate of diffusion of a gas is inversely proportional to the square root of its density.

Gram's Determinant

$$(a . b \times c)^2 \equiv [a\,b\,c]^2 = \begin{vmatrix} a . a & a . b & a . c \\ b . a & b . b & b . c \\ c . a & c . b & c . c \end{vmatrix}$$

Grashof's Number

A dimensionless number, (Gr), for flow of a fluid past a body given by

$$(Gr) = \frac{\gamma T a\, l^3 \rho^2}{\eta^2}$$

where γ is the volume coefficient of thermal expansion for the fluid, a the acceleration, ρ the density, η the dynamic viscosity and l a characteristic linear dimension of the body.

Graves' Theorem

$$f\left[\pi_1 + \frac{d\phi}{d\rho_1},\, \pi_2 + \frac{d\phi}{d\rho_2}\right] \equiv e^{\phi\,(\rho_1,\,\rho_2)} \cdot f(\pi_1,\,\pi_2) \cdot e^{-\phi\,(\rho_1\,.\,\rho_2)}$$

where π and ρ are two distributive symbols of operation, combining according to the law

$$\rho\,.\,\pi \equiv \pi\,.\,\rho + a$$

Green's Dyadic

A vector operator corresponding to **Green's Function** when the appropriate differential equation is expressed in terms of vectors.

Green's Function

The Green's function $G_k(r,r_0)$ is used to solve linear partial differential equations subject to boundary conditions.
Thus a solution of

$$\nabla^2\psi + k^2\psi = -4\pi\rho(r)$$

subject to arbitrary Dirichlet or Neumann conditions on the closed boundary surface S, may be expressed in terms of $G_k(r,r_0)$ which itself is a solution of a similar equation

$$\nabla^2 G_k(r,r_0) + k^2 G_k(r,r_0) = -4\pi\delta(r - r_0)$$

and subject to similar, but homogeneous, boundary conditions on the same surface S. Here δ is a **Kronecker's Delta Function**, r is a direction vector to the observer's point (where we require the solution) and r_0 is a direction vector to the source point (usually on S).

134

Green's Theorem

A relation expressing an integral taken over the surfaces of a number of bodies as an integral taken through the space between them.

If u, v, w are continuous functions of the Cartesian coordinates x, y, z, then

$$\sum \int\int (lu + mv + nw)\, \mathrm{d}S = -\int\int\int \left(\frac{\partial u}{\partial x} + \frac{\partial v}{\partial y} + \frac{\partial w}{\partial z}\right)\, \mathrm{d}x\, \mathrm{d}y\, \mathrm{d}z$$

where Σ denotes that the surface integrals are summed over any number of closed surfaces and l, m, n are direction cosines of the normal drawn in every case from the element $\mathrm{d}S$ into the space between the surfaces. The volume integral is taken through the space between the surfaces.

In the special case, when

$$u = U\frac{\partial V}{\partial x},\ v = U\frac{\partial V}{\partial y},\ w = U\frac{\partial V}{\partial z}$$

Green's theorem becomes

$$\int\int\int (U\nabla^2 V - V\nabla^2 U)\, \mathrm{d}x\, \mathrm{d}y\, \mathrm{d}z = -\sum \int\int \left(U\frac{\partial V}{\partial n} - V\frac{\partial U}{\partial n}\right)\mathrm{d}S$$

where $\partial/\partial n$ denotes differentiation along the normal to the surface S.

Gregory's Interpolation Formulae

These are obtained from **Newton's Interpolation Formula** when equal intervals are used. If $f(x)$ is given for $x_0, x_0 + h, \ldots x_0 + nh$, then

$$f(x) = f(x_0) + \frac{x - x_0}{h}\Delta f(x_0) + \frac{(x - x_0)(x - x_0 - h)}{2!\, h^2}\Delta^2 f(x_0)$$

$$+ \ldots + \frac{(x - x_0)\ldots\{x - x_0 - (n-1)h\}}{n!\, h^n}\Delta^n f(x_0)$$

$$+ \frac{(x - x_0)\ldots(x - x_0 - nh)}{(n + 1)!}\left(\frac{\mathrm{d}^{n+1}\, f(x)}{\mathrm{d}f(x)^{n+1}}\right)_{x=\zeta}$$

where $\Delta f(x_0) = f(x_0 + h) - f(x_0)$

$$\Delta^2 f(x_0) = \Delta f(x_0 + h) - \Delta f(x_0)\ \text{etc.,}$$

and ζ has a value intermediate between the greatest and least of x_0, $x_0 + nh$ and x. The formula is called **Gregory's Forwards Formula**. In

135

addition **Gregory's Backwards Formula** applies for $f(x)$ given for x_0, $x_0 - h, \ldots x_0 - nh$, when

$$f(x) = f(x_0) + \frac{x - x_0}{h} \nabla f(x_0) + \frac{(x - x_0)(x - x_0 + h)}{2! \, h^2} \nabla^2 f(x) \ldots$$

where $\nabla f(x_0) = f(x_0) - f(x_0 - h)$ etc.
This form is often used for extrapolation.

Gregory's Series

$$\tan^{-1} x = x - \frac{x^3}{3} + \frac{x^5}{5} - \ldots + (-1)^{n-1} \frac{x^{2n-1}}{2n-1} + \ldots$$

that value of $\tan^{-1} x$ being understood which starts from zero with x.

Grignard Reaction
The alkyl magnesium halides, Grignard reagents, are widely used in synthetic organic chemistry. The reagent is usually prepared by adding the alkyl halides to clean magnesium turnings in dry ether. The reaction may be initiated by the addition of a small crystal of iodine. The reagent is always used in the solvent in which it is prepared. The number of reactions in which a Grignard reagent can take part is very great but for brevity most of the reactions may be classified under two heads:

(a) Reaction of the reagent with a group containing multiple bonds, for example a carbonyl, nitroso, isocyanide, cyanide or sulphone group. No reaction occurs with ethylenic or acetylenic bonds. The alkyl group adds on to the group with lower electron affinity whilst the MgX radical adds into the other group, e.g.

This compound may then be hydrolysed or further treated.
(b) A Grignard reagent will react with compounds containing an active hydrogen group or a reactive halogen atom

$$R.MgX + R'OH \rightarrow RH + (R'O)MgX$$

The method may be exemplified by the preparation of a ketone.

$$R-\overset{\overset{\displaystyle O}{\|}}{C}-Cl \;+\; R'MgX \;\rightarrow\; \left[R-\overset{\overset{\displaystyle OMgX}{|}}{\underset{\underset{\displaystyle R'}{|}}{C}}-Cl \right] \;\rightarrow\; RCOR' + MgXCl$$

Grotthus and Draper Law
Only radiations absorbed by a system are effective in producing chemical change.

Grotthus' Chain Theory
To explain the conductivity of electrolytes, von Grotthus, in 1805, made the assumption that the electrodes behaved as the poles of a magnet, the cathode attracting hydrogen and the anode attracting oxygen. He made no attempt to explain the nature of the forces involved, but imagined the molecules of the electrolyte stretched out in chains between anode and cathode with decomposition taking place on the molecules nearest to these electrodes.

Grove's Process
Alkyl chlorides may be produced by passing hydrochloric acid into the alcohol in the presence of anhydrous zinc chloride.

$$C_2H_5OH + HCl \xrightarrow{\;ZnCl_2\;} C_2H_5Cl + H_2O$$

Grüneisen Constant or **Number**
The constant first appeared in the Grüneisen equation of state for solid bodies;

$$pV + G(V) = \gamma E$$

where p is the pressure, $G(V)$ is the potential energy, V is the molar volume and $E = \int_0^T C_V \, dT$ is the energy of atomic vibrations. γ is essentially a measure of the anharmonicity of the vibrations and is independent of temperature for most elements, lying between 1·5 and 2·5. For certain materials such as LiF and diamond which have high Debye temperatures γ drops with temperature.

Grüneisen Relation
The coefficient of thermal linear expansion a is related to the bulk modulus κ by the equation

$$a = \frac{\gamma \, C_V}{3 \, \kappa \, v}$$

where C_V is the specific heat at constant volume and γ is the Grüneisen Constant.

Gudden-Pohl Effect
The flash of light or luminescence which occurs in certain materials such as zinc sulphide phosphor when an electric field is applied or removed while the material is exhibiting phosphorescence (or afterglow). The effect is often called electro-photoluminescence.

Gudermannian
If $\cosh x = \sec \theta$ then θ is called the Gudermannian of x.

Guerbet Reaction
Higher branched-chain primary alcohols result when normal primary alcohols are heated with sodium ethoxide in the presence of a nickel catalyst.

$$RCH_2CH_2OH + RCH_2CH_2OH \xrightarrow[\substack{Ni \\ 250^\circ C}]{NaOC_2H_5} RCH_2CH_2\overset{\displaystyle |}{\underset{\displaystyle R}{C}}HCH_2OH + H_2O$$

Guillemin Effect
A magnetostrictive effect in which, if a bar of ferromagnetic material is elastically or permanently bent, it tends to straighten upon magnetization.

Guldberg and Waage's Law
This law, known more commonly as the Law of Mass Action, states that, for a homogeneous system, the rate of a chemical reaction is proportional to the active masses of the reacting substances. The molecular concentration of a substance in solution or in the gas phase may be taken as a measure of the active mass.

Guldin's or Guldinus' Theorems See Pappus' Theorems

Gunn Effect

The occurrence of high frequency (microwave region) current variations in certain semiconductors, notably gallium arsenide, on application of a high d.c. electric field. For a semiconductor to exhibit the effect it must possess a negative change of electron mobility with electric field once the electric field has reached a threshold value. In gallium arsenide this decrease in the mobility arises from the transfer of electrons from a high mobility central valley in the conduction band to satellite valleys which are at a higher energy but where the mobility is lower. Under these conditions the charge distribution in the semiconductor sample becomes unstable and there is established a high electric field domain which passes down the sample to the anode where it collapses. On formation of the domain the current in the sample falls, whereas on collapse of the domain the current increases again, allowing the nucleation of a new domain at the cathode. Hence a regular series of pulses is observed. This effect has technological application in the production of a simple semiconductor microwave generator.

Gurevich Effect

In an electrical conductor, over which a temperature gradient exists, phonons will themselves carry a thermal current (the lattice heat flow) and if phonon—electron collisions are important—as they are in a pure metal at low temperatures—the phonons may tend to drag the conduction electrons with them from hot to cold. This effect will produce an additional component of thermoelectric power.

H

Haber Process

The reaction between nitrogen and hydrogen to form ammonia is exothermic and accompanied by a diminution in volume

$$N_2 + 3H_2 \rightleftharpoons 2NH_3 + Q$$

By **Le Chatelier's Principle** therefore the reaction will be assisted by low temperatures and high pressures. There is, however, an optimum temperature on the practical scale, for although low temperatures will ultimately yield greater amounts of ammonia, the speed of reaction will be greatly reduced.

Hadamard's Inequality

If a determinant of order n has value D and complex elements a_{ij} then

$$\left| D \right|^2 \leqslant \prod_{i=1}^{n} \sum_{j=1}^{n} \left| a_{ij} \right|^2$$

Hagedorn and Jensen Estimation (of Sugar in the Blood)

The blood is coagulated by heating with sodium carbonate and zinc sulphate. The filtrate is heated in an alkaline solution with a measured amount of potassium ferricyanide which is partially reduced to the ferrocyanide. The liquid is treated with potassium iodide and zinc sulphate and then acidified. The excess ferricyanide liberates iodine whilst the ferrocyanide formed is precipitated as the double salt in the alkaline solution.

$$2Fe(CN)_6{}^{3-} + 2I^- \rightarrow I_2 + 2Fe(CN_6)^{4-}$$

$$2Fe(CN)_6{}^{4-} + 2K^+ + 3Zn^{2+} \rightarrow K_2Zn_3[Fe(CN)_6]_2 \downarrow$$

The iodine is titrated with thiosulphate and the difference between this value and one obtained from a blank gives the quantity of sugar in the urine.

Hagen-Rubens Relation

The reflectivity in the infrared region of a metal is given by

$$\rho = 1 - 2\left(\frac{f}{\mu \, \sigma_0} \right)^{1/2}$$

where f is the frequency of the radiation and σ_0 and μ are respectively the electrical conductivity in zero magnetic field and the magnetic permeability of the metal.

Haidinger Interference Fringes
Fringes produced by the interference of light reflected from or transmitted by two plane parallel surfaces of a thick transparent plate. The fringes are at infinity and not in the plane of the plate as in the case of very thin films.

Hall Effect
The Hall effect is a phenomenon which may be observed when conductors and semiconductors are subjected to an electric and magnetic field orthogonal to each other. A voltage, the **Hall voltage** V_H, appears across the sample in a direction at right angles to both these fields. If the current density arising from the applied electric field is j_x, the magnetic flux density is B_z and the thickness is Δy then a **Hall coefficient** R can be defined;

$$V_H = R \, j_x \, B_z \, \Delta y.$$

It can be shown for the case of electrons in a metal that the Hall coefficient is related to the carrier density n by the relationship.

$$R = -\frac{1}{ne}.$$

For electrons in a pure semiconductor, because of a different velocity distribution, this relationship becomes

$$R = -\frac{3\pi}{8} \frac{1}{ne}.$$

For holes the sign is reversed, and hence the measurement of the Hall voltage is useful practically to distinguish between n- and p-type materials. If both carriers are present as in an intrinsic semiconductor, the relationship is more complicated;

$$R = -\frac{3\pi}{8e} \cdot \frac{n\mu_n^2 - p\mu_p^2}{(n\mu_n + p\mu_p)^2}$$

where μ_n and μ_p are the Hall mobility of electrons and holes respectively and n and p are their carrier densities.

The angle between the direction of current flow and the total electric field (i.e. the vector sum of the applied field and the Hall field) is called the **Hall angle**.

Hall-Heriault Process
The production of aluminium by electrolysis of alumina dissolved in molten cryolite.

Hallwach's Effect
A negatively charged body in a vacuum is discharged by exposing it to ultraviolet light.

Hamiltonian

$$H = T + V$$

where T = kinetic and V = potential energy, is a Hamiltonian, if the total energy H can be expressed in terms of conjugate variables, i.e. momenta and coordinates only (and time if necessary).

If $T = p^2/2m$ where p the momentum and m the mass, the **Hamiltonian Operator**, as used in quantum mechanics, becomes

$$H_{op} = -\frac{\hbar^2}{2m} \nabla^2 + V$$

where V is a function of position.

Hamilton Jacobi Equation
The differential equation

$$H\left(\frac{dS}{dq}, q\right) = E$$

where H is the Hamiltonian of the system, q the generalized coordinates and S is a contact transformation function such that $dS/dq = p$ (momentum).

Hamilton's Canonical Equation
In a generalized physical system with n degrees of freedom expressed in generalized coordinates $q_1, q_2, \ldots q_n$, and generalized momenta $p_1, p_2, \ldots p_n$, and with $H = H(q_r, \dot{q}_r, t)$, the **Hamiltonian** of the system, then

$$\dot{p}_i = -\frac{\partial H}{\partial q_i} \text{ and } \dot{q}_i = \frac{\partial H}{\partial p_i}$$

($i = 1, 2, \ldots n$). *See also* **Lagrange's Equations of Motion**.

Hamilton's Principle

In a dynamical system composed of discrete material particles where the kinetic energy T of the system is known as a function of the coordinates and their derivatives and the potential energy V of the system is known as a function of coordinates and the time, then the motion of the system is such that an integral,

$$\int_{t_1}^{t_2} (T - V)(q_1, q_2, \ldots, q_n, \dot{q}_1, \dot{q}_2, \ldots, \dot{q}_n, t) \, dt$$

where the q's represent the generalized coordinates necessary to specify the configuration of the system, is as small as possible.

Hammett Equation

An equation relating reaction rate constants to structural parameters. It is essentially a linear free energy relationship.

$$\triangle\triangle G = \rho \triangle\triangle G_s$$

where $\triangle\triangle G$ represents the effect of a certain substituent on the free energy of activation, $\triangle\triangle G_s$ is the effect of that same substituent in an arbitrarily chosen standard reaction and ρ is a constant characteristic of the reaction being correlated.

Hammick-Illingworth Rule

If, in a benzene derivative C_6H_5XY, Y is in a higher group in the periodic table or if in the same group is of lower atomic weight than X then the second group to enter the nucleus goes into the m-position. In all other cases including that in which XY is a single group, the group is p-directing. When the groups joining X and Y are the same ($-C=C-$) the group XY is o-p orientating. With mixed groupings the rule may be rigidly applied. Thus for $CHCl_2$, CH will be o-p orientating and CCl will be m-directing and, as is found in practice, mixtures of the three substituents are formed.

Hankel Function

The Hankel functions are linear combinations of the ordinary solutions of Bessel's equation. Thus

$$H_n^{(1)}(x) = J_n(x) + jN_n(x)$$

$$H_n^{(2)}(x) = J_n(x) - jN_n(x)$$

where J_n and N_n are **Bessel Functions**.

143

Hankel's Integral

$$\frac{1}{\Gamma(x)} = \frac{j}{2\pi} \oint (-t)^{-x} e^{-t} \, dt$$

The path of integration C starts at $+\infty$ on the real axis, circles the origin anticlockwise and returns to the starting point.

Hansen's Integral Formula

A Bessel Function of order n can be represented by the integral equation

$$J_n = \frac{1}{2\pi} \int_{-\pi}^{\pi} e^{ix \cos t} e^{in(t - \pi/2)} \, dt$$

$$= \frac{(-1)^n}{\pi} \int_0^{\pi} e^{ix \cos t} \cos nt \, dt \qquad (n = 0, 1, 2, \ldots)$$

Hantzsch Synthesis

A pyrrole derivative is formed when chloroacetone, a β-ketoester and a primary amine condense together.

$$
\begin{array}{c}
CH_2Cl \\
| \\
CH_3 \!-\! C \!=\! O
\end{array}
+ NH_3 +
\begin{array}{c}
CH_2COO.C_2H_5 \\
| \\
O \!=\! C.CH_3
\end{array}
$$

$$
\rightarrow
\begin{array}{c}
CH \!-\!\!-\! C.COOC_2H_5 \\
\| \qquad \| \\
CCH_3 \quad CCH_3 \\
\diagdown \;\; \diagup \\
NH
\end{array}
+ HCl + 2H_2O
$$

Hardy-Schulz Rule

The flocculating power of ions increases rapidly with their charge.

Hargreave's Equation

A theorem in operational mathematics. Written symbolically:

$$\psi(D).\phi(x) \equiv e^{D.\Delta} \phi(x).\psi(D)$$

where $D \equiv \dfrac{d}{dx}$ and $\Delta \equiv \dfrac{d}{dD}$.

(*Phil. Trans., London* (1848), 31–54.)

Harnack's Theorem

Let L be a simple contour and let S^+, S^- be the finite and infinite parts into which the plane is divided by L (which does not itself belong to S^+ or S^-). Let $f(t)$ be a real and continuous function on L. Then, if

$$\frac{1}{2\pi j} \int \frac{f(t)\, dt}{t - z} = 0 \text{ for all } z \text{ in } S^+, f(t) = 0 \text{ everywhere on } L:$$

also if

$$\frac{1}{2\pi j} \int \frac{f(t)\, dt}{t - z} = 0 \text{ for all } z \text{ in } S^-, f(t) = \text{const. on } L.$$

Hartmann Dispersion Formula

An expression giving the variation of refractive index n with wavelength

$$n = n_0 + \frac{A}{(\lambda - \lambda_0)^a}$$

where n_0, A, λ_0 and a are constants for a given material.

Hartmann Test

To determine the aberration in a lens, a diaphragm containing a number of small apertures is placed before the lens and the course of the rays is plotted by photographing the pencils of light in planes on either side of the focus.

Hartree Equation

In a magnetron, the rotating field may be expressed in terms of a number of sinusoidal travelling waves rotating about the cathode with angular velocity given by

$$\omega = \frac{2mf_n}{Nm + n}$$

where f_n is the frequency of oscillation, N the number of gaps, n the mode of oscillation and m is an integer.

Hartree System

A system of units in which unity is assigned to $\hbar$, the quantum unit of angular momentum, to e, the charge of the electron in electrostatic units, and to m, the mass of the electron. The unit of length is $\hbar^2/me^2$ (the radius of the first Bohr orbit for an electron moving nonrelativistically around a proton of infinite mass) and the unit of time is $\hbar^3/me^4$ (the time for the electron in the Bohr orbit to move through one radian). *See also* **Bohr-Sommerfeld Atom.**

Haworth Synthesis
This preparation converts benzene into naphthalene.

Hay's Test (for Bile Salts)
The bile salts lower the surface tension of water considerably and thus a light powder such as flowers of sulphur will not float in a solution containing a concentration of bile salts.

Heaviside Equations
Differential equations applicable to a transmission line;

$$-\frac{\partial V}{\partial x} = Ri + L\frac{\partial i}{\partial t}$$

$$-\frac{\partial i}{\partial x} = GV + C\frac{\partial V}{\partial t}$$

where R, L, G and C are the resistance, inductance, leakage and capacitance per unit length of line, i and V are current and voltage along the line and x and t are distance and time.

Heaviside Layer
Heaviside suggested that an ionized layer in the air, concentric with the earth's surface, might serve as a reflecting surface which would

confine radiation between it and the earth. The existence of such a layer, or rather system of layers, has been conclusively proved. The main levels have been given the names 'Heaviside' or 'Kennelly-Heaviside Layer' for the lower layer and the **'Appleton Layer'** for the upper layer. More recent practice has been to call them the E and F layers respectively.

Heaviside-Lorentz System of Units
C.g.s. (centimetre, gramme, second) units in which the force between two magnetic poles m_1 and m_2 distance d apart in a medium of permeability μ is $m_1 m_2 / \mu d^2$ and in which the force between two charges q_1 and q_2 a distance r apart in a medium of permittivity ϵ is $q_1 q_2 / \epsilon r^2$.

Heaviside's Expansion Theorem
Let

$$u = \frac{\mathrm{F}(p)}{\mathrm{f}(p)} u_0$$

where $\mathrm{F}(p)$ and $\mathrm{f}(p)$ are known functions of $p = \partial/\partial t$.
  Heaviside's expansion theorem states that

$$u = u_0 \left\{ \frac{\mathrm{F}(0)}{\mathrm{f}(0)} + \sum \frac{\mathrm{F}(a)}{a\mathrm{f}'(a)} e^{|at} \right\}$$

where a is any root, except 0, of $\mathrm{f}(p) = 0$, $\mathrm{f}'(p)$ is the first derivative of $\mathrm{f}(p)$ with respect to p, and the summation extends over all the roots of $\mathrm{f}(p) = 0$.
  The solution reduces to $u = 0$ at $t = 0$.

Heaviside's Unit Function
A voltage which is zero for $t < 0$ and unity for $t \geq 0$.

Hefner *See* **Appendix**

Heisenberg Force
An exchange force between nucleons in which charge (equivalent to spin plus position) is exchanged. (*See also* **Majorana Force**.)

Heisenberg Uncertainty Principle

This principle, first put forward by W. Heisenberg in 1927, is one of the fundamental bases of modern physics. The simultaneous precise determination of velocity or any related property (e.g. momentum or energy) of a material particle and its position is impossible. The smaller the particle the greater the degree of uncertainty. (*See* **Poisson's Bracket.**)

Heitler-London Covalence Theory

Lewis proposed on empirical grounds that a covalent bond corresponds to an electron duplet. Heitler and London, generalizing from a wave-mechanical treatment of the hydrogen molecule, postulated that such an electron duplet is formed by electrons having opposite spins as it had been found by them that two hydrogen atoms can only combine when the spins of the two electrons are anti-parallel.

Heller's Test (for Albumins)

When a solution containing albumins is 'layered' on to the surface of a concentrate nitric acid solution, a white ring is formed at the junction. This is used for testing for albumins in urine.

Hell-Volhard-Zelinsky Reaction

The usual method of preparing a-chloro or a-bromo acids is to treat the acid with chlorine or bromine in the presence of red phosphorus.

$$R.CH_2.COOH \xrightarrow{P + Br_2} RCH_2COBr \rightarrow R.CH.Br.CO.Br$$

Helmert's Formula

An expression giving the value of g, the acceleration due to gravity in cm/sec^2 for a latitude l and altitude h in metres above sea level.

$$g = 980 \cdot 616 - 2 \cdot 5928 \cos 2l + 0 \cdot 0069 \cos^2 2l - 0 \cdot 0003086h$$

Helmholtz Double Layer

Electrokinetic phenomena (e.g. endosmosis) were explained by G. Quincke in 1859 a's due to a charged layer at the liquid-solid interface. This concept of the existence of differently charged layers or electrical double layers was extended by H. von Helmholtz in 1879 who suggested that a double layer is invariably formed at the boundary between two phases. If the electrical double layer is regarded as an electrical condenser of parallel plates a molecular distance apart, the subject may be treated mathematically.

Helmholtz Function (or Free Energy)

$$F = U - TS$$

where F is the Helmholtz function or free energy, U is the internal energy, S is the entropy and T the absolute temperature.

Helmholtz's Equation (for a Reversible Electrolyte Cell)
If E is the e.m.f. of a reversible cell, W the work equivalent of the chemical reaction when charge passes through the cell and T is the absolute temperature

$$E = W + T \frac{dE}{dt}$$

Helmholtz's Equation (Partial Differential Equation)
A partial differential equation

$$\nabla^2 \psi + k^2 \psi = 0$$

obtained from the wave equation when the time dependence is harmonic.

Helmholtz's Relation (Optics)
If a_1 and a_2 are the inclinations of light rays to an axis passing through a spherical surface where the refractive indices on the respective sides of the surface are n_1 and n_2, then the object and image heights y_1 and y_2 are related by

$$y_1 n_1 \tan a_1 = y_2 n_2 \tan a_2.$$

The equation is only valid for small angles and may be approximated by

$$y_1 n_1 \, a_1 = y_2 n_2 \, a_2.$$

Usually in this form it is referred to as **Lagrange's Law.**

Helmholtz's Theorem *See* **Thévénin's Theorem**

Helmholtz's Theorem (for Fluids)
In the absence of body forces (when entropy is a constant), individual vortices in a non-viscous fluid always consist of the same fluid particles.

149

Henderson Equation (for Continuous Mixture Boundaries)

The combination of an electrode of unknown potential with a reference electrode often involves contact between two solutions of different electrolytes. A liquid junction potential is introduced therefore into the observed e.m.f. of the cell. P. Henderson, from the assumption that the boundary between two different electrolytes consists of a series of mixtures of the electrolytes in all proportions, was able to deduce an equation for the potential.

$$E_B = \frac{RT}{F} \frac{c(u_+ + u_-) - c'(u_+' + u_-')}{c(u_+z_+ - u_-z_-) - c'(u_+'z_+' + u_-'z_-')} \ln \frac{c(u_+z_+ + u_-z_-)}{c'(u_+'z_+' + u_-'z_-')}$$

where c and c' are the concentrations of the two electrolytes, u and u' are the speeds of the ions and z and z' are the valencies of the ions. The subscripts refer to anions and cations.

Henderson Equation for pH

In 1908 L. J. Henderson deduced that the pH of an acid during neutralization may be represented by

$$pH = pK_a + \log \frac{[salt]}{[acid]}$$

where pK_a is the logarithm to base 10 of the reciprocal of the dissociation constant of the acid. This equation is applicable for solutions between pH 4 and pH 10, providing they are not too dilute.

Henrici's Notation See Bow's Notation

Henry's Law

The mass of gas dissolved by a given volume of solvent at a constant temperature is proportional to the pressure of gas with which the solvent is in equilibrium.

Hermann-Mauguin Symbols

A notation used in describing the symmetry classes of crystals. See Schoenflies Crystallographic Notation.

Hermite Polynomials

Solutions of the Hermite differential equation

$$\frac{d^2y}{dx^2} - 2x\frac{dy}{dx} + 2ny = 0$$

when n is an integer.

The polynomials are given by

$$H_n(x) = (2x)^n - \frac{n(n-1)}{1!}(2x)^{n-2} + \frac{n(n-1)(n-2)(n-3)}{2!}(2x)^{n-4} + \dots$$

or by the coefficients in the expansion

$$e^{2xs-s^2} = \sum_{n=0}^{\infty} \frac{H_n(x)\,s^n}{n!} \quad |s| < \infty$$

The Hermite polynomials give the probability density function (the wave function) of a simple harmonic oscillator in quantum theory.

Hermitian Matrix
If A is a square matrix, then the Hermitian adjoint A^+ is given by $(A^+)_{mn} = (A)^*_{nm}$ where rows and columns in A have been interchanged and then the complex conjugate of each element taken. A Hermitian matrix is a matrix equal to its own Hermitian adjoint; i.e. $A^+ = A$. A skew-Hermitian or anti-Hermitian is given by $A^+ = -A$.

Hermitian Operator
A quantum mechanical operator, say P, such that

$$\int_\tau u^*.Pv\;d\tau = \int_\tau v.P^*u^*\;d\tau$$

where * denotes conjugate quantity and the integration is over the whole space.

Heron's (or Hero's) Formula
The area of a triangle of sides a, b and c is given by

$$A = \sqrt{\left[s(s-a)(s-b)(s-c)\right]}$$

where $s = \frac{1}{2}(a + b + c)$.

Hertz *See* **Appendix**

151

Hertz Effect
The enhancement of a spark discharge by passing ultraviolet light to promote ionization.

Hertzian Waves
Electromagnetic waves of frequencies between zero and approximately 10^4 Mhz.

Hertz's Law
The radius of contact between a sphere and a surface is given by

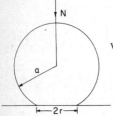

$$r = 1 \cdot 11 (Na/E)^{1/3}$$

where r is the radius of contact,
   a is the radius of the sphere,
   N is the normal force exerted on the sphere,
   E is the Young's modulus of the material of the sphere.

Hertz Vector
If, in an electromagnetic field, **A** is the vector potential, then the Hertz vector is given by

$$\Pi = \frac{c}{\mu} \int \mathbf{A} \, dt$$

where μ is the magnetic permeability of the medium.

Herzig-Meyer Method
Methyl-imino groups may be quantitatively estimated by heating the compound with hydrogen iodide at 150°C.

$$Ar.NH.CH_3 + HI \xrightarrow{150°C} Ar.NH_2 + CH_3I$$

Hessian

$$\text{If } y_k = \frac{\partial F}{\partial x_k} \quad (k = 1, 2, \ldots, n)$$

where F is a function of $x_1, x_2, x_3, \ldots, x_n$, the symmetrical determinant,

$$H = \left| \quad \frac{\partial^2 F}{\partial x_i, \partial x_j} \quad \right|$$

is the Hessian.

152

Hess' Law
The heat change in a chemical reaction is the same whether the reaction takes place in one or several stages.

Heurlinger Equations
Equations relating adiabatic and isothermal thermoelectric and thermomagnetic effects. (In an adiabatic effect there is no transverse flow of heat whereas in an isothermal effect there is no transverse temperature gradient.) Thus if subscripts a and i represent the adiabatic and isothermal coefficients respectively;

$$R_a = R_i + \Sigma_i P$$
$$Q_a = Q_i - \Sigma_i S$$
$$\rho_a = \rho_i - Q_i P B^2$$
$$\lambda_a = \lambda_i (1 + S^2 B^2)$$
$$\Sigma_a = \Sigma_i + Q_i S B^2$$

where R is the Hall coefficient, P the Ettingshausen coefficient, Σ the thermoelectric power, ρ the electrical resistivity, λ the thermal conductivity and B the magnetic flux density. (See also the **Hall, Ettingshausen, Nernst-Ettingshausen, Righi-Leduc** and **Maggi-Righi-Leduc Effects.**)

Heusler Alloys
Ternary metal alloys which contain manganese, aluminium and copper but no ferromagnetic element but which exhibit ferromagnetism. Other similar ternary ferromagnetic alloys also exist.

Hilbert Transform
If

$$H(z) = \frac{1}{\pi} \mathscr{P} \int_{-\infty}^{\infty} \frac{h(y)}{y - z} \, dy$$

then

$$h(y) = -\frac{1}{\pi} \mathscr{P} \int_{-\infty}^{\infty} \frac{H(z)}{z - y} \, dz$$

are Hilbert transforms of each other. $\mathscr{P}$ indicates that the principal parts of the integrals should be taken.

153

Hill's Determinant
A determinant occurring in the solution of **Mathieu's Equation**.
The zeros of this determinant establish the stability of the solution.

Hiltner-Hall Effect
The polarization of light from distant stars. It is thought to occur in interstellar regions and to be due to the selective absorption of light vibrating in one plane by ferromagnetic grains which are aligned in a magnetic field in these regions.

Hinsberg Separation
A mixture of the three amines can be separated into its components by means of p-toluene sulphonyl chloride, $CH_3.C_6H_4.SO_2Cl$. After treatment with this reagent the solution is made alkaline with potassium hydroxide. The three amines react in the following ways.

Primary Amines:
$$RNH_2 + ClO_2SC_6H_4CH_3 \rightarrow RNHO_2SC_6H_4CH_3 + HCl$$

The complex is soluble in potassium hydroxide

$$RNH.O_2SC_6H_4CH_3 \xrightarrow{KOH} RNKO_2SC_6H_4CH_3 + H_2O$$

Secondary Amines:

$$\begin{array}{c} R \\ \diagdown \\ NH \\ \diagup \\ R' \end{array} + Cl.O_2.SC_6H_4CH_3 \rightarrow \begin{array}{c} R \\ \diagdown \\ NO_2SC_6H_4CH_3 \\ \diagup \\ R' \end{array} + HCl$$

The complex is insoluble in potassium hydroxide.

Tertiary Amines:

$$\begin{array}{c} R \\ \diagdown \\ R \rightarrow N \\ \diagup \\ R \end{array}$$ cannot react with the p-toluene sulphonyl chloride.

The mixture is now steam distilled to separate out the tertiary amine, then the secondary amine sulphonyl chloride is filtered off and the primary amine sulphonyl chloride is regenerated from the filtrate by the action of hydrochloric acid. The amines may be regenerated by refluxing with 70% sulphuric acid or 25% hydrochloric acid.

Hoesch Reaction

If a nitrile is used instead of hydrogen cyanide in the **Gatterman Aldehyde Synthesis,** a ketone is formed. This modification is known as the Hoesch reaction.

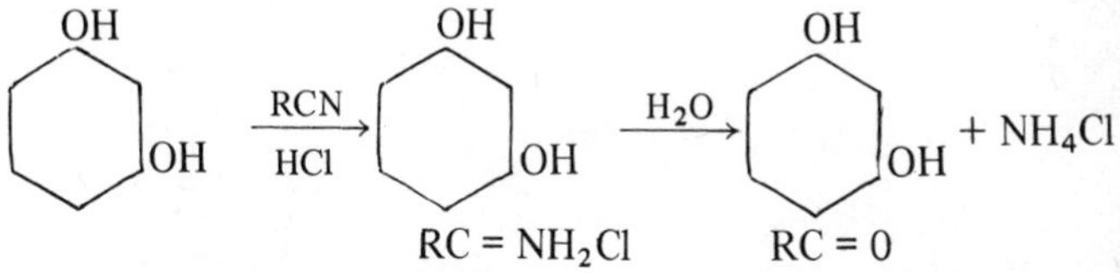

$$RC = NH_2Cl \qquad RC = 0$$

Hofmann Amine Separation

The mixture of the three amines is heated with ethyl oxalate. Tertiary amines do not react, primary and secondary amines react according to the following schemes.

I.

$$
\begin{array}{l}
COOC_2H_5 \\
| \qquad\qquad + 2RNH_2 \rightarrow \\
COOC_2H_5
\end{array}
\qquad
\begin{array}{l}
COONHR \\
| \qquad\qquad + 2C_2H_5OH \\
COONHR \\
\text{Crystalline}
\end{array}
$$

II.

$$
\begin{array}{l}
COOC_2H_5 \\
| \qquad\qquad + R_2NH \rightarrow \\
COOC_2H_5
\end{array}
\qquad
\begin{array}{l}
COONR_2 \\
| \qquad\qquad + C_2H_5OH \\
COOC_2H_5 \\
\text{Liquid}
\end{array}
$$

The reaction mixture is then distilled. The first fraction contains the tertiary amine, the second fraction the diamide, whilst the residue in the distillation flask contains the monoamide. The amines themselves are regenerated from the amides by hydrolysing with alkali and distilling off the amine.

Hofmann Degradation

This reaction results in the conversion of an amide into a primary amine with one carbon atom less by the action of bromine or fluorine and an alkali. It may be assumed to take place by the following mechanism.

155

$$RCNH_2 \xrightarrow{\quad Br \quad} RCNHBr$$

$$\begin{array}{ccc}
\underset{\displaystyle O}{\overset{\displaystyle \|}{RCNH_2}} & \xrightarrow{\ Br\ } & \underset{\displaystyle O}{\overset{\displaystyle \|}{RCNHBr}} \\
& & \downarrow KOH \\
\left[\underset{\displaystyle O}{\overset{\displaystyle \|}{RCN}}\right] & \xleftarrow{-KBr} & \underset{\displaystyle O}{\overset{\displaystyle \|}{RCNKBr}} \\
\downarrow & & \\
RNCO & \xrightarrow{KOH} RNH_2 &
\end{array}$$

This reaction can be carried out using the amides of all mono-carboxylic acids and with yields of up to 90% for acids having up to seven carbon atoms.

Hofmann Exhaustive Methylation Reaction

The thermal decomposition of quaternary ammonium compounds is a most important reaction. Tetramethylammonium hydroxide is the only compound of this type which decomposes to give an alcohol.

$$(CH_3)_4NOH \rightarrow (CH_3)_3N + CH_3OH$$

All other quaternary ammonium halides give an olefin and water.

$$(C_2H_5)_4NOH \rightarrow (C_2H_5)_3N + H_2O + C_2H_4$$

If the compound contains different alkyl groups, one of which is methyl, this group remains attached to the nitrogen. Further, if an ethyl group is present, ethylene is formed preferentially to any other olefin (**Hofmann Rule**). Water is preferentially eliminated between the hydroxyl group and a β hydrogen atom in one of the alkyl groups. This reaction is the basis of the Hofmann exhaustive methylation reaction which is used to determine the position of a nitrogen atom in a heterocyclic ring rather than to prepare unsaturated compounds.

Hofmann Mustard Oil Reaction

Alkyl isothiocyanates may be prepared by heating together a primary amine, carbon disulphide and mercuric chloride. The mechanism of the reaction is uncertain but it may occur by the intermediate formation of the dithiocarbamic acid salt.

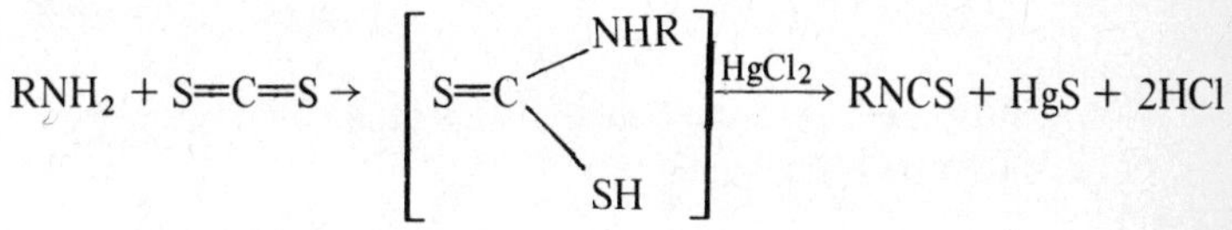

$$RNH_2 + S{=}C{=}S \rightarrow \left[S{=}C \underset{SH}{\overset{NHR}{\diagdown\diagup}} \right] \xrightarrow{HgCl_2} RNCS + HgS + 2HCl$$

Hofmann Rearrangement

This reaction may be used to prepare the homologues of aniline. For example, when phenyl trimethyl ammonium chloride is heated under pressure, the following reaction occurs.

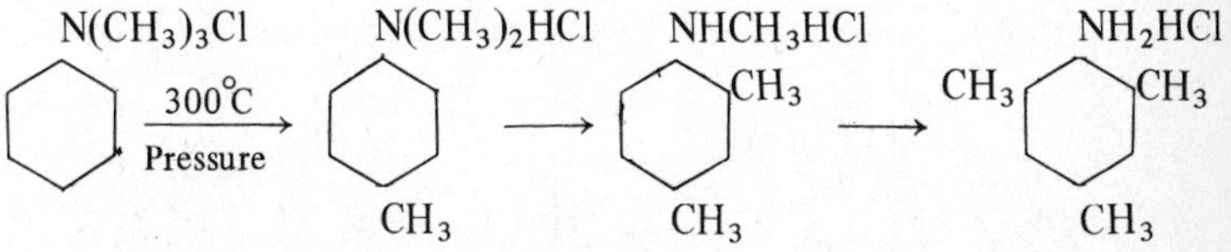

Hofmann Rule *See* **Hofmann Exhaustive Methylation Reaction**

Hofmeister Series

A list of anions and cations arranged in the order of their increasing effectiveness in causing the swelling of gelatine.

Hölder Condition

Let $f(t) = f_1(t) + jf_2(i)$ be some complex function of the point t of L. If, for every pair of points t_1 and t_2 of L

$$|f(t_2) - f(t_1)| \leqslant A|t_2 - t_1|^\mu$$

where A and μ are positive constants, $f(t)$ is said to satisfy the Hölder condition. A and μ are called the **Hölder constant** and **index**.

Hölder's Inequality

If f_1 and f_2 are two functions

$$\left| \int_a^b f_1 f_2 \, dx \right| \leqslant \left[\int_a^b |f_1|^p \, dx \right]^{1/p} \left[\int_a^b |f_2|^{p/(p-1)} \, dx \right]^{(p-1)/p}$$

$$(1 \leqslant p < \infty)$$

157

The inequality implies a similar inequality for sums of convergent series;

$$\left| \sum_j a_j b_j \right| \leqslant \left[\sum_j |a_j|^p \right]^{1/p} \left[\sum_j |b_j|^{p/(p-1)} \right]^{(p-1)/p}$$

The inequality reduces to the **Cauchy-Schwarz Inequality** if $p = 1$. See also **Minkowski's Inequality**.

Hooke's Law
Within the elastic limit of any body, the ratio of stress to strain is constant.

Hopkins-Cole (Glyoxylic Acid) Reaction
A violet ring obtained when concentrated sulphuric acid is added to a mixture containing a protein and glyoxylic acid.

The tryptophan nucleus

is believed to condense with the aldehyde to form a coloured product. Exceptions to this test are gelatine and zein, a corn protein.

Hopkinson's Coefficient
The ratio of the mean flux per turn of an induction winding to the mean flux per turn of another winding linked with it.

Hopkins Reaction for Lactic Acid
An alcoholic solution of lactic acid is heated with concentrated sulphuric acid and 3 drops of copper sulphate. The mixture is warmed in a beaker of boiling water for 5 minutes, cooled under the tap and 2 drops of a 0·2% alcoholic solution of thiophen is added. After rewarming in boiling water a cherry-red coloration develops.

Horner's Method (for Determining the Real Roots of an Equation)
Let

$$f(x) = a_0 x^n + a_1 x^{n-1} + \ldots a_n x + a_n = 0$$

be the equation and let p_1 be a first approximation to the root x_1

sought, such that $p_1 + 1 > x_1 > p_1$. Diminish the roots of the equation by p_1. In the transformed equation

$$a_0(x - p_1)^n + a_1(x - p_1)^{n-1} + \ldots + a_{n-1}(x - p_1) + a_n = 0$$

put

$$\frac{p_2}{10} = \frac{a_n}{a_{n-1}}$$

and diminish the roots by $p_2/10$ yielding

$$b_0\left(x - p_1 - \frac{p_2}{10}\right)^n + b_1\left(x - p_1 - \frac{p_2}{10}\right)^{n-1} + \ldots + b_n = 0.$$

Then take

$$\frac{p_3}{100} = \frac{b_n}{b_{n-1}}$$

and continue the operation. The required root will be

$$x_1 = p_1 + \frac{p_2}{10} + \frac{p_3}{100} + \ldots$$

If, however, b_n and b_{n-1} are of the same sign, p_2 was taken too large and must be diminished.

Houben-Hoesch Synthesis
This preparation is an extension of the **Gatterman Aldehyde Synthesis** but is not applicable to phenol itself. Cyanides condense with polyhydric phenols (particulary m-compounds) in the presence of zinc chloride and hydrogen chloride to give phenolic ketones.

Hubble Effect
Distant galaxies show, in their spectra, red shifts which increase proportionally with the distance of the galaxy. This effect is consistent with the hypothesis of an expanding universe.

Hugoniot Function *See* Rankine-Hugoniot Relations

Hume-Rothery Rules
Hume-Rothery pointed out that particular alloy phases often occur at the same ratio of valence electrons to atoms.

Phase	Electrons/Atoms
a (face-centred cubic)	1·36–1·42
β (body-centred cubic)	1·48–1·50
γ (complex)	1·58–1.67

Hund's Rule
A rule enabling the relative order to be obtained of the different energy states for a configuration of equivalent electrons; i.e. when the electrons have the same quantum numbers n (principal quantum number) and l (orbital angular momentum) and thus form a sub-shell in an atom. Hund's rule states that the configuration which has the largest multiplicity or S value (where S is the quantum number for resultant spin of the electrons) is the lowest in energy, and if there are several of these terms, then the lowest of these has the highest value of L (quantum number for resultant orbital momentum).

Hunsdieker Reaction
If a silver carboxylate is allowed to react with an equivalent quantity of bromine in carbon tetrachloride at boiling point the alkyl halide is produced. The reaction provides a method of reducing the length of a carbon chain by one carbon atom.

$$RCOOAg + Br_2 \rightarrow RBr + CO_2 + AgBr$$

Hurtley's Test (for Acetoacetic Acid in Urine)
To 10 c.c. of urine add 2·5 c.c. of concentrated hydrochloric acid and 1 c.c. of a freshly prepared 1% solution of sodium nitrite. Shake and allow to stand for 2 minutes. Add 15 c.c. of 0·880 ammonia and 5 c.c. of 10% ferrous sulphate. Shake, pour into a large boiling tube and allow

160

to stand. A violet or purple colour develops if acetoacetic acid is present.

Hüttig Equation
An equation used to measure the surface area of non-porous solids by gas adsorption

$$\frac{V}{V_m} = \frac{c^{p/p_s}}{1 + c^{p/p_s}} \left(1 + \frac{p}{p_s} \right)$$

where p = equilibrium gas pressure at temperature T,
$\quad p_s$ = saturated vapour pressure of the adsorbate at the temperature of adsorption,
$\quad c$ = $A \exp\{ (q-q_L)/RT \}$
$\quad A$ = constant,
$\quad q$ = heat of adsorption into a first layer molecule,
$\quad q_L$= heat of liquefaction of the adsorbate,
$\quad V$ = volume of gas adsorbed at p and T.
$\quad V_m$ = volume of gas required to cover the surface completely with a unimolecular layer.

Huygens' Principle. (Huygens-Fresnel Principle)
Every point of a wave can be considered as the centre of a new elementary wave. The resultant wave produced by the interference of all these elementary waves is identical with the original wave.

Hylleraas Coordinate
A coordinate system which leads to a separate solution of Schrödinger's equation for the two particle system, e.g. Helium atom. The coordinates are $v = r_{12}, s = r_1 + r_2, t = r_1 - r_2$, where r_1 and r_2 are the distances of two electrons from the nucleus and r_{12} is the inter-electron distance.

I

Ilkovik Equation

In polarography, the limiting diffusion current I_d in amps. of a substance reduced or oxidized at the dropping-mercury electrode, is given by

$$I_d = 0 \cdot 63 Z F c D^{1/2} w^{2/3} t^{1/6}$$

where Z is the valency,

       F is the Faraday

       D is the diffusion coefficient of the substance,

       w is the weight of mercury dropping per second,

       t is the time between drops in seconds,

       c is the concentration of the reacting species.

Ising Model

Used, for instance, in ferromagnetism and phase transformation theory, it is a model for a system of interacting particles. A scalar variable σ associated with each lattice point of a crystal has possible values $+1$ or -1 representing the two directions of electron spin in a ferromagnet or the two types of atoms in a binary alloy. The interaction energy is given by

$$E_{ij} \begin{cases} = -J\sigma_i\sigma_j \text{ where } i, j \text{ are nearest neighbour particles} \\ = 0 \qquad \text{where } i, j \text{ are not nearest neighbours.} \end{cases}$$

J is a coupling constant which can be positive as in a ferromagnetic system or negative as in an antiferromagnetic system. The model enables a value for the heat capacity to be calculated; the two-dimensional form and the three-dimensional form (which has not been solved exactly) lead to a transition point.

Ivanov Reagent

A reagent analogous to a Grignard reagent (*see* **Grignard Reaction**);

$$C_6H_5.CH.COOMgBr.$$
$$|$$
$$MgBr$$

It reacts with carbon dioxide to give phenyl malonic acid, iodine to give substituted succinic acids and with aldehydes and ketones to produce a phenyl β hydroxy acids.

J

Jacobian
If

$$y_k = f_k(x_1, x_2, x_3, \ldots, x_n)$$

the determinant:

$$J = \begin{vmatrix} \dfrac{\partial y_1}{\partial x_1} & \dfrac{\partial y_1}{\partial x_2} & \cdots & \dfrac{\partial y_1}{\partial x_n} \\[2ex] \dfrac{\partial y_2}{\partial x_1} & \dfrac{\partial y_2}{\partial x_2} & \cdots & \dfrac{\partial y_2}{\partial x_n} \\[1ex] \cdots\cdots\cdots\cdots\cdots \\[1ex] \dfrac{\partial y_n}{\partial x_1} & \dfrac{\partial y_n}{\partial x_2} & \cdots & \dfrac{\partial y_n}{\partial x_n} \end{vmatrix}$$

is the Jacobian.

Jacobi Polynomials
The Jacobi polynomials are defined as

$$J_n\,(a, c\,; x) = F\,(a + n, -n; c; x)$$

where F is the hypergeometric function and n is an integer. They are solutions of the differential equation

$$(x^2 - y^2)\frac{d^2y}{dx^2} + \left[(1 + a)x - c\right]\frac{dy}{dx} + n(a + n)y = 0$$

See also **Gauss' Hypergeometric Differential Equation.**

Jacobi's Elliptic Functions *See* **Legendre's Elliptic Integrals**

Jacobi's Identity

$$\Big[X, [Y, Z] \Big] + \Big[Y, [Z, X] \Big] + \Big[Z, [X, Y] \Big] = 0$$

where [ ] are **Poisson Brackets**.

Jacobi's Theta Functions

If ν is a complex variable and $q = e^{i\pi\tau}$ is a complex parameter such that τ has a positive imaginary part then the four theta functions are given by

$$\vartheta_1(\nu) = 2 \sum_{n=0}^{\infty} (-1)^n q^{(n+\frac{1}{2})^2} \sin\{(2n+1)\pi\nu\}$$

$$\vartheta_2(\nu) = 2 \sum_{n=0}^{\infty} q^{(n+\frac{1}{2})^2} \cos\{(2n+1)\pi\nu\}$$

$$\vartheta_3(\nu) = 1 + 2 \sum_{n=1}^{\infty} q^{n^2} \cos(2n\pi\nu)$$

$$\vartheta_4(\nu) = 1 + 2 \sum_{n=1}^{\infty} (-1)^{n^2} \cos(2n\pi\nu)$$

Jacobsen Rearrangement

During sulphonation, polyalkylbenzenes or halogenated polyalkyl benzenes fairly readily undergo isomerization, due to the migration of the alkyl group or the halogen. The reaction may be intra- or inter-molecular and both migrations have been shown to occur at the same time. The alkyl group migrates to the vicinal position.

This reaction has not been found to occur in compounds containing $-NH_2$, $-NO_2$, $-OCH_3$ or $-COOH$.

165

Jaffé's Test for Indican
A test for indican which depends upon the blue coloration of a chloroform layer when a solution containing indican is treated with a large excess of hydrochloric acid and 12% of potassium chlorate.

Jaffé's Test of Creatinine
Treat 10 c.c. of the solution with 15 c.c. of saturated picric acid solution and 5 c.c. of 10% caustic soda. Allow the solution to stand for 5 minutes and dilute to 200 c.c. A deep orange colour is developed due to the formation of a tautomeric form of creatinine picrate.

Jahn Teller Effect
The electron states of a non-linear polyatomic molecule cannot show degeneracy in the ground state; i.e. there cannot be more than one possible quantum mechanical state of the same energy. Thus, if degeneracy of the electron states due to symmetry of the molecule should exist, the atoms in the molecule will be displaced to destroy this symmetry.

Japp Klingemann Reaction
Compounds containing an active merlylene group e.g. acetoacetic ester couple with diazonium salts. The reaction was used to detect the presence of the enolic form in keto-enol mixtures. The mechanism of the reaction is believed to be,

$$C_6H_5 - N = N^+ + \overline{C}H\,(COMe)\,CO_2\,C_2H_5 \xrightarrow{NaCH}$$

$$C_6H_5 - N = N - CH\,(COMe)\,CO_2\,C_2H_5 \longrightarrow$$

$$C_6H_5 - NH - N = C \begin{array}{c} COMe \\ \diagup \\ \diagdown \\ CO_2\,C_2H_5 \end{array}$$

Jeans Viscosity Equation
An equation relating the viscosity η of a gas to the temperature:

$$\eta = kT^n$$

where n is an empirical constant.

Johnsen-Rahbek Effect

If a piece of semiconducting material is placed in contact with two electrodes, the frictional force between the electrodes and the semiconductor increases with increase of applied voltage to the electrodes. The effect is additional to any increase of friction that may occur due to electrostatic attraction of the electrodes. Similar variation of the shear resistance of a fluid with variation of voltage applied to two electrodes immersed in the fluid is called the **Winslow Effect**.

Johnson Noise

In electronics, noise arising from thermal agitation of electrons in conductors.

Jones Symbols

A notation for expressing symmetry operations within a point group. The symbols express the new positions of the x, y and z axes after the operation. For instance, a $180°$ rotation about the x axis is expressed by the symbol $\bar{x}\,\bar{y}\,\bar{z}$.

Jones Zone

This is similar and in certain cases equal to the **Brillouin Zone** for a solid and the distinction is not always made. It is a zone for which an energy gap exists at all points on its surface. The surface of the zone generally consists of partial Brillouin zone boundaries from several zones.

Josephson Effect

If two superconductors are connected by a thin layer of insulating material with no electrostatic potential present, there is a quantum mechanical penetration, through the layer, of electrons existing as **Cooper Pairs**. The arrangement is called a Josephson junction and the tunnelling is known as the **d.c. Josephson Effect**.

If a steady potential difference V is applied, an oscillatory current passes through the layer of frequency

$$\omega = 2\pi\left(\frac{e'\,V}{h}\right)$$

where e' is the effective charge of the Cooper pairs. This is known as the a.c. **Josephson Effect**.

167

Joshi Effect

A fall (negative effect) or rise (positive effect) in the low-frequency alternating current passing through a gas dielectric condenser when the gas is irradiated continuously with visible light.

Joule *See* Joule's Law, *also* Appendix

Joule Effect

A rod of para- or diamagnetic material suffers a mechanical change in length when subjected to a magnetic field. *See also* Joule's Law.

Joule's Law

The heating effect of electric currents is in accordance with the conservation of energy. The unit of work in the derived system of electromagnetic units is the **Joule** and is 10^7 ergs. It is the work done on one coulomb moving through a difference of potential of one volt.

Joule-Thompson Coefficient *See* Joule-Thompson Effect

Joule-Thompson Effect (Joule-Kelvin Effect)

When a gas is allowed to expand adiabatically through a porous plug the temperature of the gas changes. The rate of change of temperature with pressure on adiabatic expansion i.e.

$$\left(\frac{\partial T}{\partial p}\right)_H$$

known as the **Joule-Thompson Coefficient** may be taken to be constant if the pressure difference across the plug is small. The differential Joule-Thompson effect, $\Delta T/\Delta p$, where ΔT is the change in temperature for a small change in pressure Δp at temperature T can be shown from thermodynamical considerations to be given by the expression

$$\Delta T = \frac{T[\partial V/\partial T]_p - V}{C_p} \Delta p$$

where V is the volume of the gas and C_p is the specific heat at constant pressure. The sign of the Joule-Thompson Coefficient will depend upon the relative values of the terms in the numerator of the above expression. The value of the numerator is dependent upon the temperature and there is a fixed temperature at which the sign of the Joule-Thompson Coefficient changes. This temperature is known as the **Joule-Thompson inversion temperature.** Most gases are below their inversion temperatures at room temperature and hence may be liquefied by expansion. The so-called permanent gases (e.g. hydrogen, $T_i = 190°K$), have an inversion temperature below room temperature and must be cooled before they can be liquefied by this means.

Jurin Rule
The height of rise of a liquid in a capillary tube of radius a is given by:

$$h = \frac{2\gamma \cos \alpha}{\rho g a}$$

where γ is the surface tension of the liquid, α is the angle of contact with the capillary and ρ is the density of the liquid.

K

Kaiser Effect
Very low energy acoustic pulses, which are observed when a metal is initially deformed, and which are not observed on reapplication of any stress up to the original limit. The effect arises from work-hardening removing the plastic region of the stress-strain curve. Acoustic emission reappears upon reversal of the stress or upon application of a stress exceeding the original limit. *See also the* **Bauchinger Effect**.

Kapp Line *See* Appendix

Kayser *See* Appendix

Kekulé Benzene Formula
Kekulé in 1865 assigned a ring structure to benzene which he wrote as

Ladenburg in 1869 pointed out that the Kekulé structure should give rise to two *o*-derivatives and proposed a three-dimensional structure, viz.:

Structures of this type, however, were proved to be incorrect when Ladenberg in 1874 showed that the six hydrogen atoms are all equivalent—as the Kekulé formula tacitly assumes, but this formula did not account for the fact that the double bonds in benzene were not of the same nature as the ethylenic double bond. Claus in 1867 introduced his formula

to get over this difficulty and it was assumed that *p*-bonds were not like ordinary bonds but could be easily ruptured. Dewar postulated a quinonoid structure for benzene but as it was asymmetrical it was discarded.

Kekulé in 1872 modified his views to suggest that benzene was a tautomeric mixture.

Some support for this view was forthcoming much later (Levine and Cole 1932) from the ozonolysis products of *o*-xylene; i.e. glyoxal, methyl glyoxal and dimethyl glyoxal. Various other structures were proposed, e.g. the Armstrong-Baeyer Centric Formula

which postulated the existence of bonds encountered nowhere else in the realm of organic chemistry.

Modern thought favours the theory that benzene is a resonance hybrid of the following structures

The Kekulé structures contribute 80% to the resonance hybrid whilst the Dewar structures only contribute 20%. The resonance energy is 37 kcal/mole. Bond measurements show that C–C bonds

are 1·54 Å, C=C bonds are 1·33 Å and the benzene bond is 1·39 Å and thus has a length between that of a double and single bond.

Kellogg Rule
An equation relating the pressure, absolute temperature and density ρ of a gas, of the form:

$$p = RT\rho + (B_0RT - A_0 - C_0/T^2)\rho^2 + (bRT - a - c/T^2)\rho^3$$

Kelvin Effect *See* Thomson Effect

Kelvin's Equation
This relates the Peltier coefficient to the thermoelectric power for a conductor (or semiconductor).

$$\pi\,(B) = T\,\Sigma\,(B)$$

where $\pi\,(B)$ is the Peltier coefficient which is a function of magnetic flux density B if the conductor is placed in a magnetic field, and $\Sigma\,(B)$ is the thermoelectric power. (*See also* **Peltier** *and* **Nernst Effects**.)

Kelvin Skin Effect
A non-uniform distribution of variable currents in solid conductors resulting in an increase in current density near the surface.

Kelvin's Statement (of the Second Law of Thermodynamics)
No process is possible whose sole result is the abstraction of heat from a reservoir and the performance of an equivalent amount of work; i.e. it is impossible to construct a perfect heat engine. (*See also* **Clausius' Statement**.)

Kelvin's Theorem
For an ideal homogeneous fluid in a region of flow where the entropy is constant the circulation around a closed path moving with the fluid remains constant; i.e. vorticity cannot be created or destroyed under these conditions.

172

Kelvin Temperature Scale

Practical temperature scales usually depend upon the expansion of a particular substance and since the coefficients of expansion are not constant a scale defined in this fashion will be subject to some uncertainty. W. Thompson (Lord Kelvin) suggested a temperature scale based on the efficiency of a reversible machine. The zero of the Kelvin scale is the temperature of the sink of such a machine working at unit efficiency, i.e. converting heat completely into work. This is only possible at absolute zero on the gas scale of temperature and hence, provided that the degree is defined in the same fashion, these two scales are identical. By definition since 1954, $0°K$ is $-273 \cdot 16°C$.

Kennard Packet

The minimum uncertainty state function $\chi(\mathbf{r})$ which can be used to describe a particle. It is given by

$$\chi(\mathbf{r}) = [2\pi(\Delta\mathbf{r})^2]^{-\frac{1}{4}} \exp\left[\frac{-(\mathbf{r} - <\mathbf{r}>)^2}{4(\Delta\mathbf{r})^2}\right]$$

Thus the particle is centred at position $<\mathbf{r}>$ with uncertainty $\Delta\mathbf{r}$. The corresponding uncertainty in momentum space is obtained by changing all $\mathbf{r}$ to $\mathbf{p}$ where $\mathbf{p}$ is the momentum. Any arbitrary decrease in $\Delta\mathbf{r}$ would increase $\Delta\mathbf{p}$ causing the wave packet to spread out again with time.

Kennedy's Theorem

Any three bodies having plane motion relative to each other have only three *centros* which lie on a straight line. A *centro* is either a point common to two bodies having the same velocity, or a point on a body about which another body turns or tends to turn.

Kepler's Equation

An expression for motion in an ellipse;

$$\theta - \epsilon \sin\theta = \frac{2\pi t}{T}$$

θ is the eccentric angle, ϵ the eccentricity and T the period.

173

Kepler's Laws

I. Every planet moves in an ellipse of which the sun occupies one focus.

II. The radius vector drawn from the sun to the planet sweeps out equal areas in equal times.

III. The squares of the times taken to describe their orbits by two planets are proportional to the cubes of the major semiaxes of the orbits.

Kerr Electro-optic Effect

Dr. John Kerr discovered in 1875 that certain materials when placed in a stationary electric field become optically anisotropic and doubly refracting. If the electric field is applied perpendicular to the direction of the light then the effect is called the Kerr effect. If n_1 and n_2 are the refractive indices for light whose planes of polarization are respectively parallel and perpendicular to the electric field E, then

$$n_1 - n_2 = k \lambda E^2$$

where λ is the wavelength of the light and k is Kerr's constant. *See also* **Pockel's Effect.**

Kerr Magneto-optic Effect

Light, plane polarized in or normal to the plane of incidence, becomes elliptically polarized when reflected from the polished pole of an electromagnet.

Ketteler-Helmholtz Formula

In media which exhibit anomalous dispersion, the refractive index n for light of wavelength λ is given by

$$n^2 = 1 + \sum \frac{A_m \lambda^2}{\lambda^2 - \lambda_m^2}$$

if the medium contains sets of molecules with different periods of vibration, where λ_m is the wavelength in free space of light of the same frequency as the natural frequency of the mth type of molecule, A_m is a constant referring to that molecule and the summation is taken to include all the natural periods of vibration of the molecules in the substance.

Khintchine's Theorem

If $x_1, x_2, \ldots x_n$ is a sequence of statistically independent random variables which all have the same probability distribution with mean value ζ, then as $n \to \infty$

$$x = \frac{1}{n}(x_1 + x_2 + \ldots x_n)$$

converges in probability to ζ.

Kick's Law

In crushing material

$$E = k \ln (d_1/d_2)$$

where E = energy used in crushing,

k = constant depending on material and type of crusher,

d_1, d_2 = average linear dimensions before and after crushing.

Kikuchi Lines

Black or white lines which appear in the electron diffraction pattern of highly perfect crystals. When an electron beam is incident on the crystal some electrons are diffusely scattered with loss of energy. These electrons then can be reflected back in the direction of the plate used to exhibit the diffraction pattern. A series of these processes gives rise to white (weakened) lines and black (enhanced) lines on the plate, these lines being parallel to and equidistant from the projection of each crystal plane.

Kiliani Reaction

The sugar series may be ascended by this reaction. Thus if an aldopentose is treated with hydrocyanic acid, a cyanhydrin is formed which on hydrolysis with barium hydroxide and acidification with the correct quantity of dilute sulphuric acid gives rise to a poly-hydroxyacid. The solution is evaporated to dryness and the γ-lactone is formed which, on reduction with sodium amalgam in slightly acid solution, yields the aldohexose.

Kirchhoff's Approximation

A solution of the integral equation for scattering problems involving the substitution of the incident wave for the unknown scattering density function.

Kirchhoff's Equation for Heat of Sublimation

An equation of the form

$$L_s = \int_0^T C_p^f dT - \int_0^T C_p^i dT + L_0$$

where C_p^i and C_p^f are the specific heats at constant pressure at the initial and final states and L_0 is the heat of sublimation at absolute zero.

Kirchhoff's Equations

The relationship between the variation of heat of reaction ΔH with temperature and the change of heat capacity ΔC_p at constant pressure p during a process is given by

$$\left[\frac{\partial(\Delta H)}{\partial T}\right]_p = \Delta C_p.$$

Similarly, if U is the energy change

$$\left[\frac{\partial(\Delta U)}{\partial T}\right]_V = \Delta C_V.$$

Kirchhoff's Law of Emission

The emissive power divided by the absorption coefficient for any substance depends only on the frequency and plane of polarization of the radiation and the temperature and is independent of the nature of the substance.

Kirchhoff's Laws of Current

(a)

$$I_1 + I_2 = I_3 + I_4 + I_5$$

(b)

$$I_1 R_1 + I_2 R_2 + I_3 R_3 + I_4 R_4 - I_5 R_5 - I_6 R_6 = 0$$

(c)

$$E_1 = I_2 R_2 + I_3 R_3 + I_4 R_4 + I_5 R_5$$

Kirchhoff's Laws of Current

In any circuit composed of a network of conductors:

(1) The algebraic sum of the currents at any junction of conductors (branch point) must be zero. $\Sigma\, I = 0$.

(2) $\Sigma\, IR = \Sigma\, E$. The left-hand side sums the products of the currents in, and the resistances of, parts of the conductors forming a closed circuit in the network and the right-hand side is the sum of all the discontinuities of potential met in passing round the circuit. The proper signs must be given to the I's and E's throughout.

Kirkendall Effect

If inert marker wires are embedded at the common face of a copper and a brass (CuZn) block, which are held at a high temperature, the wires marking the original interface move into the brass. This is because the zinc atoms diffuse past the interface faster than the copper atoms. On condensing, excess vacancies can cause the formation of voids in the brass. The effect, discovered by F. C. Kirkendall (1942) and confirmed in the above experiment by Smigelskas and Kirkendall (1947) and since shown in many other systems, provides proof of the predominance of a vacancy mechanism for diffusion in pure metals and substitutional alloys.

Klein-Gordon Equation

An equation of the form

$$\frac{1}{c^2}\frac{\partial^2 \psi}{\partial xt^2} = \frac{\partial^2 \psi}{\partial x^2} - \mu^2 \psi$$

which describes the motion of a flexible string embedded in an elastic medium. In quantum mechanics it can be used to describe the 'scalar' meson, i.e. one without a spin. In the presence of spin this equation becomes the **Proca Equation.**

Klein-Nishina Formula

The scattering cross-section per electron for Compton scattering into solid angle $d\Omega$ is given for unpolarized radiation by

$$d\sigma_C = \frac{r_0^2}{2}\cdot\left(\frac{E_p'}{E_p}\right)^2\cdot\left(\frac{E_p}{E_p'} + \frac{E_p'}{E_p} - \sin^2\theta\right)\cdot d\Omega$$

where r_0 is the classical radius of the electron.
For remaining notation *see* **Compton Effect.**

177

Knight Shift

W. D. Knight (1949) discovered that the nuclear magnetic resonance frequency in metals (or alloys) is higher than it is in chemical compounds containing nuclei of the same isotopes in the same magnetic field. The effect is due to the small extra field at the nucleus produced by electrons in the conduction band.

Knoevenagel Reaction

If an aldehyde is treated with an equivalent amount of diethyl-malonate in the presence of pyridine and the product hydrolysed and heated, an α-β unsaturated acid is formed.

$$R.C\begin{array}{c}H\\\\O\end{array} + H_2C\begin{array}{c}COOC_2H_5\\\\COOC_2H_5\end{array} \xrightarrow[pyridine]{} R.C=C\begin{array}{c}H\\COOC_2H_5\\\\COOC_2H_5\end{array}$$

$$\xrightarrow[heat]{hydrolyse} RCH=CH.COOH + 2C_2H_5OH + CO_2$$

In practice it is usually sufficient to treat the aldehyde with malonic acid in the presence of pyridine.

Knoop's β-Oxidation Theory

Compounds containing an aromatic nucleus are not so readily oxidized by organisms as those containing straight chains. Knoop combined the phenyl group with various fatty acids, fed the compounds formed and investigated the products eliminated in the urine. He found that no matter what the acid, the product was either hippuric acid or phenylaceturic acid.

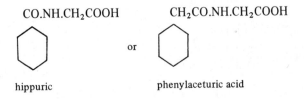

$CO.NH.CH_2COOH$

hippuric

or

$CH_2CO.NH.CH_2COOH$

phenylaceturic acid

These results indicated in his view that oxidation of the fatty acids occurred in such a way that at each stage in the degradation process there was a loss of two carbon atoms due to oxidation of the β carbon.

Thus, no matter from what acid we start, i.e. $C_6H_5(CH_2)_nCOOH$, we must arrive at hippuric acid when n is even and phenylaceturic acid when n is odd. This hypothesis has been confirmed by Schonheimer and Rittenberg who isolated deuteropalmitic acid (C_{16}) from the body fat of rats fed on deuterostearic acid (C_{18}).

Knorr Synthesis
Pyrrole derivatives may be prepared by the condensation of an a-amino ketone with a β-diketone or β-ketoester

Knudsen Cosine Law
Kinetic theory shows that for a gas at rest and at a constant temperature the number of molecules dn striking or leaving an area dA of the wall subtending a solid angle dϕ at an angle θ to the normal is given by

$$dn = \frac{dA}{4\pi} nc \cos \theta \ d\phi$$

where n is the number of molecules per unit volume and $\bar{c}$ is the mean velocity of the molecules.

Knudsen Flow
That flow which occurs when the mean free path of the molecules is greater than the dimensions of the apparatus. The ratio of the latter to the former is often called **Knudsen's number**.

Knudsen Formula for Gas Flow
The mass of gas q passing any point in a tube per second at a certain temperature and very low pressure is given by

$$q = \frac{(2\pi\rho)^{\frac{1}{2}}}{6} d^3 \frac{p_1 - p_2}{l}$$

where $(p_1 - p_2)/l$ is the pressure gradient along the length of the tube l, d is the diameter of the tube and ρ is the density of the gas at the given temperature and unit pressure.

Kohlrausch Law (of Independent Migration of Ions)

Each ion contributes a definite amount to the total conductance of the electrolyte irrespective of the nature of the other ion. This rule is strictly true only at infinite dilution where there is no ionic attraction between the ions, i.e.

$$\Lambda_0 = \lambda_+ + \lambda_-$$

where λ_+ and λ_- are the ionic conductances at infinite dilution and Λ_0 is the equivalent conductivity.

Kohlrausch Square Root Law

In the calculation of equivalent conductance of electrolytes, the extrapolation of Λ_C, the equivalent conductance at concentration C gm mol per litre, to Λ_0, the equivalent conductance at zero concentration, is made easier by plotting Λ_C as ordinate against $\sqrt{C}$ as abscissa, the relationship being linear.

Kolbe Reaction

Most aliphatic acid salts upon electrolysis with a smooth platinum electrode give rise to a hydrocarbon and carbon dioxide at the anode.

$$2CH_3COO' \rightarrow C_2H_6 + 2CO_2$$

Kolbe Synthesis

This is the original method of preparing salicylic acid. If sodium phenoxide is heated with carbon dioxide at 180–200° the following reaction occurs.

This reaction was modified by Schmitt (**Kolbe-Schmitt Synthesis**).

Sodium phenoxide is heated at 120–140°C under pressure with carbon dioxide and totally converted to the sodium salt of salicylic acid.

Konowaloff's Rule
The vapour from a mixture of two liquids is relatively richer in the component whose addition to a liquid mixture results in an increase of total pressure.

Kopp's Law
The specific heat of a solid element is the same whether free or combined.

Korner's Absolute Method (for Assigning the Position of Substituents in Benzene)
This method is based on the principle that the introduction of a third substituent into a *p*-isomer gives one trisubstituted product; into an *o*-isomer two such products and into a *m*-isomer three tri-substituted isomers.

In practice this method is difficult as the separation of the isomers may be almost impossible when present in minute amounts. The reverse reaction is better. Thus if each of the six diamino-benzoic acids is heated with soda lime, three phenylenediamines are formed. Three of the acids give the same diamine which must therefore have the *m*-structure whereas the diamine given by two of the acids must be the *o*-isomer and the other diamine must have the *p*-structure.

Kossel Lines

These are analogous to **Kikutchi Lines** in electron diffraction and occur when a point source of radiation is used. The x-rays meet the atomic planes at a continuous range of angles and those striking at the Bragg angle (see **Bragg's Law**) are reflected. There will be a cone of such rays and these give rise to a black arc (Kossel line) on the photographic film, either for back reflection or transmission depending on the angle of the cone. Corresponding to the cone of diffracted rays there will be also a deficiency cone which can give rise to a white arc in the transmission pattern.

Kossel-Sommerfeld Law

The arc spectrum of an element having an odd atomic number shows even multiplicity whereas the arc spectrum of an element having an even atomic number shows odd multiplicity.

Köster Effect

Plastic deformation of a material lowers the elastic modulus by introducing free dislocations.

Kraemer-Sarnon Test

A test for the melting point of bitumen wherein the melting point is defined as the temperature at which mercury falls through a standard sample of bitumen heated under standard conditions.

Kramer's Theorem

In electron paramagnetic resonance in solid materials, a two-fold spin degeneracy is not removed by electric fields or spin-orbit coupling if the spin is half integral.

Kronecker Delta

An operator defined as follows:

$$\delta_{nm} = 1 \ (m = n)$$

$$= 0 \ (m \neq n)$$

Kronecker Product
In group theory, the Kronecker (or direct) product of two three-dimensional representations of a group is obtained by multiplying the characters in the two representations. The character of a representation is the sum of the elements down the leading diagonals of the matrices of the representation.

Kronig-Penney Model
An approximate one-dimensional method of obtaining features of electron propagation in a crystal by considering the crystal as having a periodic square-well type of variation of electric potential. The potential is assumed, in the limit, to be a periodic delta function of infinite height and zero width but finite area, and the method obtains discrete ranges of energy for the electrons.

Kuhn-Thomas-Reiche f-sum Rule
For an atom whose electrons are undergoing all possible transitions from (or to) one level denoted subscript 2,

$$\sum_1 f_{12} + \sum_3 f_{23} = Z$$

where subscripts 1 and 3 refer respectively to levels (including the continuum) below and above the level denoted 2, the f's are the oscillator strengths for each transition and Z is the number of optical electrons.

Kummer's Transformation
If $s = \sum_{k=0}^{\infty} a_k$ and $S = \sum_{k=0}^{\infty} b_k$ are two convergent series such that $L = \lim_{n \to \infty} (a_n/b_n) \neq 0$, then

$$s = LS + \sum_{n=0}^{\infty} \left(1 - L\frac{b_k}{a_k}\right) a_k$$

Kundt's Rule
The refractive index of a medium on the shorter wavelength side of an absorption band is abnormally low and on the longer wavelength side

abnormally high. The phenomenon of anomalous dispersion is due to this effect.

Kurie Plot (for Beta Decay)

In 1934 E. Fermi derived for the energy distribution of beta particles emitted from radioactive isotopes a formula of the form

$$\sqrt{\left[\frac{N(p)}{p^2 F(Z, p)} \right]} = C(E_{max} - E)$$

where $N(p)$ is the number of particles having momentum between p and $p + dp$, $F(Z, p)$ is a correction factor for the Coulombic interactions between the nuclear particles and the beta particles, E is the energy of the beta particles, which can be emitted with a maximum energy E_{max}, C is a constant and Z is the atomic number of the product nucleus.

The distribution can be plotted in 'straight-line' form when the graph is called a Kurie Plot or **Fermi Plot.**

L

Ladenburg Benzene Formula *See* **Kekulé Benzene Formula**

Lagrange's Differential Equation
An equation of the form

$$y = xf_1\left(\frac{dy}{dx}\right) + f_2\left(\frac{dy}{dx}\right)$$

(alternatively called **D'Alembert's Differential Equation**). It is solved
by differentiating with respect to x and treating dy/dx as the
independent variable. The general solution is given parametrically in
terms of dy/dx, whereas the singular solution is given by

$$y = xf_1(m) + f_2(m)$$

where m is the root of

$$f_1(m) - m = 0.$$

Lagrange's Equations of Motion
Consider a general physical system with n degrees of freedom expressed
in generalized coordinates $q_1, q_2, \ldots, q_n$.

Let $V = V(q_r, t)$ be the potential energy of the system, if conservative.

Let $T = T(q_r, \dot{q}_r, t)$ be the kinetic energy of the system ($\dot{q}_r = dq_r/dt$).
Then the system can be described by the set of Lagrange's equations
of state

$$\frac{d}{dt}\frac{\partial T}{\partial \dot{q}_i} - \frac{\partial T}{\partial q_i} = -\frac{\partial V}{\partial q_i} \quad (i = 1, 2, \ldots, n)$$

Lagrange's Identity
If a_1, b_1, c_1, a_2, b_2 and c_2 are any numbers then

$$(a_1 b_2 - a_2 b_1)^2 + (b_1 c_2 - b_2 c_1)^2 + (c_1 a_2 - c_2 a_1)^2$$
$$\equiv (a_1^2 + b_1^2 + c_1^2)(a_2^2 + b_2^2 + c_2^2) - (a_1 a_2 + b_1 b_2 + c_1 c_2)^2$$

Lagrange's Interpolation Formula

If $f(x)$ is given for $x = x_0, x_1, x_2 \ldots x_n$, then the function

$$g(x) = f(x_0) \frac{(x - x_1)(x - x_2) \ldots (x - x_n)}{(x_0 - x_1)(x_0 - x_2) \ldots (x_0 - x_n)} + \ldots$$

$$+ f(x_n) \frac{(x - x_0)(x - x_1) \ldots (x - x_{n-1})}{(x_n - x_0)(x_n - x_1) \ldots (x_n - x_{n-1})}$$

tends to $f(x_0)$ at $x = x_0$, $f(x_1)$ at $x = x_1$ etc. It is not especially useful as, for interpolation, it is not usually necessary to take account of all the given values of x throughout the entire range; $g(x)$ will usually be determined mainly by the adjacent tabulated values. However, most interpolation formulae are derived from Lagrange's formula.

Lagrange's Law (Optics) *See* Helmholtz's Relation

Lagrange's Multipliers

To find stationary points of $z = f(x, y)$ subject to the condition $g(x, y) = 0$, three equations

$$\frac{\partial f}{\partial x} + \lambda \frac{\partial g}{\partial x} = 0$$

$$\frac{\partial f}{\partial y} + \lambda \frac{\partial g}{\partial y} = 0$$

$$g = 0$$

are solved simultaneously giving Lagrange's Multiplier λ, and the stationary point, x, y.

Lagrange's Theorem

If $y = z + x \phi(y)$,

$$f(y) = f(z) + x \phi(z) f'(z) + \frac{x^2}{2!} \frac{d}{dz} [\{\phi(z)\}^2 f'(z)] + \ldots$$

$$+ \frac{x^n}{n!} \frac{d^{n-1}}{dz^{n-1}} [\{\phi(z)\}^n f'(z)] + \ldots$$

186

Lagrange's Theorem in Group Theory

The order of a subgroup is a factor of the order of the group.

Lagrange's Theorem of Divisibility

If p is a prime, and r is any number less than $p - 1$, the sum of the products of the numbers $1, 2, 3, \ldots, p - 1$ taken r together is divisible by p.

Lagrangian

The Lagrangian

$$L = T - V$$

sometimes called kinetic potential, is the difference between the kinetic and potential energy of a system expressed in generalized coordinates (cf. **Lagrange's Equation of Motion**). The function is usually used to express a system possessing a finite number of degrees of freedom. For a continuous system, for example a vibrating elastic solid, the Lagrangian is expressed as an integral over the coordinates x, y and z, which are continuous variables. For every value of x, y and z there is a generalized coordinate $\eta(x, y, z, t)$ and the Lagrangian is given by

$$L = \iiint \mathcal{L} \, dx \, dy \, dz$$

where

$$\mathcal{L} = \mathcal{L}\left(\frac{d\eta}{dx}, \frac{d\eta}{dy}, \frac{d\eta}{dz}, \dot{\eta}, x, y, z, t\right)$$

is called the **Lagrangian Density Function**.

Laguerre Polynomial

A solution of Laguerre's differential equation

$$x \frac{d^2 y}{dx^2} + (1 - x)\frac{dy}{dx} + ny = 0$$

where n is an integer, in the form

$$y = L_n(x) = e^x \frac{d^n}{dx^n}\left(x^n e^{-x}\right)$$

In addition, the associated (or generalized) Laguerre polynomial given by

$$y = L_n^k(x) = \frac{d^k}{dx^k} L_n (x)$$

is a solution of the equation

$$x \frac{d^2 y}{dx^2} + (k + 1 - x) \frac{dy}{dx} + (n - k)y = 0$$

Lambert   *See* **Appendix**

Lambert's Law of Absorption
If the thickness of an absorbing medium increases in arithmetic progression, the light transmitted decreases in geometrical progression.

$$I = I_0 e^{-kx}$$

where I is the intensity of light after passage through a layer of thickness x and I_0 is the intensity of the incident light. This is also known as **Bouguer's Law.**

Lambert's Law of Emission (Lambert's Cosine Law)
The intensity of the light emitted from a perfectly diffusing surface is proportional to the cosine of the angle between the normal to the surface and the direction of observation.

Lamb Shift (or Lamb-Retherford Shift)
Spectroscopists found a separation between the main components of the H_α line in the hydrogen spectrum to be less than predicted by theory of Dirac. This was shown by Lamb using quantum electrodynamic theory to be due to the $2S_{1/2}$ term (quantum numbers $l = 1, j = \frac{1}{2}$) being 0.03528 cm^{-1} higher than the $2P_{1/2}$ term (quantum numbers $l = 0, j = \frac{1}{2}$), this difference being called the Lamb shift.

Lamé's Elastic Constants
The elastic properties of an isotropic material are completely determined by two constants called Lamé's elastic constants and

denoted by μ and λ. μ is the shear modulus or rigidity coefficient (also denoted G). λ is not a named modulus but is related to the bulk modulus, K, by

$$K = \lambda + \tfrac{2}{3}\mu$$

Lamé's Functions
Lamé's equation

$$\frac{d^2z}{dx^2} = \{m(m+1)\,\wp(x) + B\}z$$

where m is a positive integer and $\wp(x)$ is the Weierstrassian elliptic function, has $2m+1$ values of B for which the equation has a solution of one or the other of the four species of ellipsoidal harmonics. If, when such a solution is expanded in descending powers of the argument, the coefficient of the leading term is taken as unity, the function so obtained is termed a Lamé function of degree m, of the first kind and of the species corresponding to the ellipsoidal harmonic taken.

Lami's Theorem
If three forces act on a particle in equilibrium, each is proportional to the sine of the angle between the other two.

Landau Fluctuations
Experimental determination of the energy loss of a fast particle frequently involves measurement of the energy lost in a thin detector in which the particles suffer many ionizing collisions but do not lose much of their initial energy. There are marked fluctuations in the observed rate of energy loss. This variation may be attributed to the **Poisson Distribution** of ionizing events and to the wide range of energies which can be lost by fast particles in an inelastic collision.

Landé Splitting Factor
Landé, partly empirically, derived a formula for g, the fraction by which the magnetic moment of an atom must be multiplied to allow for spin, in terms of the three quantum vectors, J the total momentum vector, L the azimuthal quantum vector and S the moment of electron spin,

$$g = 1 + \frac{J(J+1) + S(S+1) - L(L+1)}{2J(J+1)}$$

189

Landholt Fringe
If a brilliant source of light is viewed through two Nicol prisms orientated with their principal axes perpendicular to one another, the field is not quite dark when exact adjustment is made. The darkened field is crossed by a black fringe, which changes position if the prisms are rotated slightly. Lippich showed that this fringe was due to the fact that the directions of vibrations in various parts of the field were not parallel.

Lane's Law
If a star contracts, its internal temperature must rise. The law is independent of the source of energy but assumes that the material of the star behaves as a perfect gas.

Langevin-Debye Equation
For susceptibility in substances where the molecules possess a permanent dipole moment and where $pE \gg kT$ (see **Langevin Function**), the total polarizibility is given by

$$a = a_0 + \frac{p^2}{3kT}$$

where a_0 is the sum of the electronic and ionic contributions.

Langevin Diamagnetism Equation
The diamagnetic susceptibility per gram mole for a diamagnetic substance is given by

$$\chi_m = -\frac{Z\,e^2 N_A}{6mc^2}<r^2>$$

where N_A is Avogadro's number, Z the atomic number and $<r^2>$ the mean square distance of the electrons from the nucleus, assuming spherical symmetry.

Langevin Function
Defined by $L(x) = \coth x - (1/x)$. The function is used for the calculation of magnetization in paramagnetic substances and the electric polarization in substances containing molecules with a

permanent electric dipole moment. For the former, magnetization per mole is given by

$$M = N_A \, \mu \mathrm{L}(x) \text{ where } x = \frac{\mu H}{kT}.$$

μ is the magnetic moment per molecule and H is the magnetic field. Electric polarization is obtained by replacing μ by p, the electric dipole moment and H by E, the electric field, in the above equations. The function applies strictly to gases, where the molecules are sufficiently far apart for their mutual interactions to be negligible. (*See also* **Brillouin Function**.)

Langley *See* **Appendix**

Langmuir Adsorption Isotherm
From a consideration of the number of gas molecules striking and leaving the surface of the absorbent, and assuming the adsorbed layer to be monomolecular, Langmuir showed that the fraction of surface covered is

$$\frac{V}{V_m} = \frac{kp}{1 + kp}$$

where V is the volume of gas adsorbed, V_m is the volume of gas necessary to cover the whole surface, k is a constant for a system and p is the pressure of the gas.

Langmuir-Child Law *See* **Child's Law**

Laplace's Coefficients
Laplace's equation $\nabla^2 V = 0$ becomes, in spherical coordinates,

$$r \frac{\partial^2}{\partial r^2} (rV) + \frac{1}{\sin \theta} \frac{\partial}{\partial \theta} \left(\sin \theta \, \frac{\partial V}{\partial \theta} \right) + \frac{1}{\sin^2 \theta} \frac{\partial^2 V}{\partial \phi^2} = 0$$

Particular solutions of this equation are

$$r^m \, Y_m(\cos \theta, \phi) \text{ and } \frac{1}{r^{m+1}} \, Y_m(\cos \theta, \phi)$$

where m is a positive integer and

$Y_m(\cos \theta, \phi)$

$$= A_0 P_m(\cos \theta) + \sum_{n=1}^{n=m} \{ A_n \cos n\phi \, P_m^n(\cos \theta) + B_n \sin n\phi \, P_m^n(\cos \theta) \}$$

where

$$P_m^n(\cos \theta) = \sin^n \theta \, \frac{d^n P_m(\cos \theta)}{d(\cos \theta)^n}$$

is a Laplace coefficient or a Surface Spherical Harmonic of the mth degree and $P_m(\cos \theta)$ is a **Legendre Coefficient**.

Laplace's Equation

If, within a closed surface, there exists no charge; the potential V on the surface is given by

$$\nabla^2 V = \frac{\partial^2 V}{\partial x^2} + \frac{\partial^2 V}{\partial y^2} + \frac{\partial^2 V}{\partial z^2} = 0$$

(*see* **Poisson's Equation**).

In other words, where no electrification exists within a surface, there can be no concentration of potential on the surface.

Laplace's Equation (for the Velocity of Sound)

$$v = \sqrt{\frac{\gamma p}{\rho}}$$

where v = velocity of sound, γ is the ratio of the specific heats of the gas, p is the pressure and ρ the density.

Laplace's Integrals

$$P_m(x) = \frac{1}{\pi} \int_0^\pi \{x + \sqrt{(x^2 - 1)} \cos \phi\}^m \, d\phi$$

$$Q_m(x) = \frac{1}{\pi} \int_0^\infty \frac{d\phi}{\{x + \sqrt{(x^2 - 1)} \cosh \phi\}^{m+1}}$$

$P_m(x)$ and $Q_m(x)$ are **Legendre's Coefficients**.

192

Laplace's Law *See* **Ampère's Law**

Laplace's Principle
An irrationality in a function cannot be removed by differentiation. Conversely, no irrationality which does not occur in an integrand can appear after integration.

Laplace's Theory of Chemical Combination
Laplace suggested that the constituents of a chemical compound remained united until the forces existing between them were disturbed by the presence of a third substance for which the affinity of one of the constituents was greater than its affinity for the other constituent. He communicated this theory to Lavoisier who confirmed it experimentally.

Laplace's Transform
If $f(x) = 0$ for $x < 0$, if $\displaystyle\int_0^\infty |f(x)|^2\, e^{-2ax}\, dx$ is finite for $a > \beta$ and if

$F(p) = \displaystyle\int_0^\infty f(x)\, e^{-px}\, dx$, then $F(p)$ is called the Laplace Transform of $f(x)$ and

$$f(x) = \frac{1}{2\pi j} \int_{-j\infty+a}^{+j\infty+a} F(p)\, e^{px}\, dx \quad (a > \beta; x > 0)$$

Laplacian Operator
The Laplacian operator ∇^2 is defined by $\nabla^2 \equiv (\nabla.\nabla)$. For rectangular Cartesian coordinates:

$$\nabla^2 \equiv \frac{\partial^2}{\partial x^2} + \frac{\partial^2}{\partial y^2} + \frac{\partial^2}{\partial z^2}$$

Laporte's Rule
For dipole radiation, only transitions between terms of opposite parity are allowed; i.e. transitions between terms having either $\Sigma_i l_i$ even or having $\Sigma_i l_i$ odd are allowed where $\Sigma_i l_i$ is the summation of the absolute value of the quantum number l for all the electrons.

Larmor Precession

The action of a magnetic field on the orbital motion of an electron was stated by Larmor. He showed that a uniform field does not affect the form and inclination of the orbit to the magnetic lines of force or the frequency and velocity of the electron in its orbit. The orbit itself, however, rotates (precesses) in a uniform fashion about an axis parallel to the field and the frequency of precession is called the Larmor frequency.

Larmor's Formula

According to Larmor a single moving electron radiates energy at the rate of

$$\frac{2}{3}\frac{e^2}{c^3}\frac{dv}{dt}$$

where v is the velocity of the electron.

Laue Equations

In the diffraction of x-rays in crystals, when the diffraction is produced by a three-dimensional grating, three conditions giving the position of spectra must be satisfied. Let the three axes have lengths a, b, c, the cosines of the angles between them and the incident ray-direction be a_0, β_0, γ_0, and the corresponding cosines for the diffracted ray be a, β, γ. The conditions for diffraction—the Laue equations—are

$$a(a - a_0) = h\lambda$$
$$b(\beta - \beta_0) = k\lambda$$
$$c(\gamma - \gamma_0) = l\lambda$$

where the triple set of integers h, k, l, denotes the order of the spectrum and λ is the wavelength.

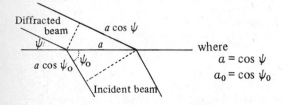

where
$$a = \cos \psi$$
$$a_0 = \cos \psi_0$$

Laue Symmetry Groups *See* Friedel's Law

Laurent's Expansion

Suppose the point a is a regular point or a pole or an isolated essential singularity of $f(z)$, but that $f(z)$ is analytic within the annules between two concentric circles whose centre is a and whose radii are r_1 and r_2. Then, within the region R

$$f(z) = \sum_{n=-\infty}^{n=+\infty} C_n(z-a)^n$$

where

$$C_n = \frac{1}{2\pi j} \int_C \frac{f(z)\,dz}{(z-a)^{n+1}} \quad (n = 0, \pm 1, \pm 2, \ldots)$$

Laves Phases

Intermediate alloy phases having the general formula AB_2 and possessing one of three related crystal structures; $MgCu_2$ type (cubic), $MgZn_2$ type or $MgNi_2$ type (both hexagonal).

Leblanc Process (for the Manufacture of Sodium Carbonate)

This process in no longer in use in Britain, having given way to the ammonia-soda or Solvay process and the electrolytic processes.

The process occurred in two parts. Common salt was heated with sulphuric acid. The sodium sulphate formed was ground and mixed with its own weight of chalk and half its weight of coal or coke and fused in a rotating furnace. The mixture of carbonate and sulphide was extracted with water and the carbonate crystallized out.

Le Chatelier's Principle

If a change occurs in one of the conditions of a system in equilibrium the system will adjust itself so as to annul, as far as possible, the effect of that change.

Lederer-Manasse Reaction

Phenol condenses with aliphatic and aromatic aldehydes in the o- and p-positions. The most important application of this reaction is the condensation of 40% formaldehyde in the presence of dilute alkali or acid.

The main product is the p-isomer and the reaction is the primary step in the production of phenol-formaldehyde resins.

Leduc Effect *Also called the* Righi-Leduc Effect

Legendre's Coefficients
The general solution of Legendre's Equation

$$(1 - x^2)\frac{d^2z}{dx^2} - 2x\frac{dz}{dx} + m(m + 1)z = 0$$

where m is a positive integer, is

$$z = A\,P_m(x) + B\,Q_m(x)$$

$$P_m(x) = \frac{(2m - 1)(2m - 3)\ldots 1}{m!}\left\{x^m - \frac{m(m - 1)}{2(2m - 1)}x^{m-2}\right.$$

$$\left. + \frac{m(m - 1)(m - 2)(m - 3)}{2.4.(2m - 1)(2m - 3)}x^{m-4} - \cdots\right\}$$

is a finite sum terminating with the term which involves x if m is odd or $x^0 = 1$ if m is even, and is called a Legendre Coefficient or a Surface Zonal Harmonic. $Q_m(x)$, a similar sum, which however has a logarithmic singularity at $x = 0$, is called a Surface Zonal Harmonic of the second kind.

Legendre's Elliptic Integrals
In the problem of the rectification of the ellipse, the two integrals

$$F(k, \phi) = \int_0^\phi \frac{d\phi}{\sqrt{(1 - k^2 \sin^2 \phi)}} = u \quad (0 < k < 1)$$

and

$$E(k, \phi) = \int_0^\phi \sqrt{(1 - k^2 \sin^2 \phi)}\, d\phi \quad (0 < k < 1)$$

arise. These are known as Legendre's forms of the elliptic integrals of the first and second kind.

The elliptic integral of the third kind may be written

$$\Pi(n, k, x) = \int_0^\phi \frac{1}{(1 + n \sin^2 \phi)\sqrt{(1 - k^2 \sin^2 \phi)}}\, d\phi \quad (0 < k < 1)$$

where n is a constant.

ϕ is defined as the amplitude of u to the modulus k

$$\phi \qquad\qquad = \text{am } u$$
$$\sin \phi \qquad\qquad = \sin \text{ am } u = \text{sn } u$$
$$\cos \phi \qquad\qquad = \text{cn } u$$
$$\sqrt{(1 - k^2 \sin^2 \phi)} = \text{dn } u$$

and sn u, cn u and dn u are **Jacobi's Elliptic Functions.**

Leibnitz's Monads
In Liebnitz's philosophy, all things, spiritual and physical, were considered in terms of fundamental entities which he called monads. As far as their physical significance was concerned, the monads were 'the very atoms of nature' but they had neither mass, parts, extension nor figure. They existed solely as centres of force or of energy.

Leibnitz's Test (for Alternating Series)
If $a_n \to 0$ monotonically, then the series $\Sigma (-1)^n a_n$ is convergent.

Leibnitz's Theorem
If u and v are functions of x

$$\frac{d^n(uv)}{dx^n} = u\frac{d^n v}{dx^n} + \frac{n}{1!}\frac{du}{dx}\cdot\frac{d^{n-1}v}{dx^{n-1}} + \frac{n(n-1)}{2!}\frac{d^2u}{dx^2}\cdot\frac{d^{n-2}v}{dx^{n-2}} + \ldots + \frac{d^nu}{dx^n}v.$$

Leidenfrost's Phenomenon
If liquid is dropped on to a hot surface, there is a critical temperature above which wetting by the liquid does not occur. The liquid becomes insulated from the surface by a layer of vapour. As a result, water, for example, when dropped on to a hot plane, breaks into globules which move erratically across the surface.

Lennard-Jones Potential
The total potential energy $\Phi(r)$ of two atoms of an inert gas at separation r is

$$\Phi(r) = \frac{B}{r^{12}} - \frac{C}{r^6}$$

197

where B and C are empirical parameters. The r^{-6} term represents the long-range attractive force (*see* **Van der Waals-London Interaction**) and the r^{-12} term the short-range repulsive force.

Lenz's Law
If an e.m.f. is induced in a circuit due to a change in the flux of magnetic induction through the circuit, the sign of the e.m.f. is such that any current flow is in a direction that would oppose the flux change.

Leuckart Reaction
Aldehydes and ketones are converted into the formyl derivatives of the corresponding amine by excess ammonium formate or formamide.

$$\text{>C=O} + 2HCOONH_4 \rightarrow \text{>CH.NH.CHO} + 2H_2O + CO_2 + NH_3$$

$$\text{>C=O} + 2HCONH_2 \rightarrow \text{>CH.NH.CHO} + CO_2 + NH_3$$

Lewis Equation *See* **Nernst-Lindemann Equation**

Lewis Number
A dimensionless characteristic number (Le) for flow phenomena involving both heat and mass transfer. It is given by

$$(Le) = \frac{\lambda}{\rho \, C \, D}$$

where λ is thermal conductivity, ρ density, C specific heat and D the coefficient of diffusion.

Lewis-Randall Rule
At a given temperature the fugacity of a constituent of a mixture of gases is proportional to its mole fraction. Up to 25 atmospheres this rule appears to hold and is a great improvement on **Dalton's Law of Partial Pressures**.

Lewis' Theory
Atoms can combine by sharing electrons, thus completing their shells without ionization.

L'Hôpital's Rule
If $f(x + L)$ and $F(x + L)$ can be both developed in Taylor's series (*see* **Taylor's Theorem**), and if $f(x) = 0 = F(x)$, the limit of the value of $f(x)/F(x)$ is $f'(x)/F'(x)$ provided this has a definite value (zero, finite or infinite). If not, the limit of the value is that of the first ratio of the derivatives that is definite.

Liebermann-Burchard Test
A colorimetric test for cholesterol. A bluish green coloration whose intensity varies with the quantity of cholesterol present is formed when a chloroform solution of the sterol is treated with acetic anhydride and concentrated sulphuric acid.

Liebermann's Nitroso-Reaction
When secondary amines react with nitrous acid, oily nitrosamines are formed

$$R_2NH + HO.NO \rightarrow R_2N.NO + H_2O$$

This procedure is used as a test for secondary amines. If a nitrosamine is warmed with a crystal of phenol and a small quantity of concentrated sulphuric acid a green solution is formed which turns blue on the addition of an alkali.

Liénard-Wiechert Potentials
The scalar and vector potentials ϕ and $\mathbf{A}$ respectively at a distance r from a moving charge q;

$$\phi\,(\mathbf{r}, t) = \frac{q}{c\,[r - \mathbf{r}\,.\,\mathbf{v}/c]_{\text{ret.}}}$$

$$\mathbf{A}\,(\mathbf{r}, t) = \frac{q\mathbf{v}}{c\,[r - \mathbf{r}\,.\,\mathbf{v}/c]_{\text{ret.}}}$$

where $\mathbf{r}$ and $\mathbf{v}$ inside the square brackets are the position and velocity at the retarded time $t' = (t - r/c)$.

Liesegang Rings
It was observed by R. E. Liesegang (1896) that when a drop of concentrated silver nitrate was placed on a film of gelatin containing

potassium dichromate, silver chromate was not precipitated as a continuous film but in a series of well defined rings. Many other instances of rhythmic precipitates are known. These rings may be obtained in solution, i.e. in the absence of a gel, providing convection currents in the solution are prevented.

Linde's Rule
An expression for the increase per atomic per cent impurity in resistivity for substitutional impurities introduced into a monovalent metal;

$$\Delta\rho = a + bz^2$$

where a and b are constants for a given row of the periodic table and a given solvent metal, and z is the excess valency (i.e. the impurity has valency $z + 1$).

Liouville's Theorem (for a function)
If a function of a complex variable $f(z)$ has no singular point for z (finite or infinite) then $f(z)$ is a constant.

Liouville's Theorem (Statistical Mechanics)
All elements of equal volume in phase space have equal *a priori* probabilities.

Lipschitz Condition
If $| f(\zeta) - f(x) | \leqslant A | \zeta - x |^a$ for given x and all $| \zeta - x | < \delta$ where A and a are independent of ζ, $a > 0$ and δ can be arbitrarily small, $f(\zeta)$ is said to satisfy a Lipschitz condition of order a at $\zeta = x$. If $f(\zeta)$ satisfies a Lipschitz condition it is continuous at $\zeta = x$, and if it satisfies one for all x in $a \leqslant x \leqslant b$, then it is continuous within this closed interval.

Lissajous Figure
When a point undergoes two periodic motions, which at any instant of time are at right angles to one another, the resultant movement of the point traces out a curve which is called a Lissajous Figure.

Lobry de Bruyn-van Ekenstein Rearrangement
In the presence of dilute alkalis or amines, polyhydric aldehydes and in particular sugars, undergo rearrangement. Thus if glucose is heated in

the presence of sodium hydroxide, an inactive solution is obtained from which D(+) glucose, D(+) mannose and D(−) fructose have been isolated. The same products are obtained if the starting material is D(−) fructose or D(+) mannose. Possibly the rearrangement takes place through the 1:2 enedrol.

$$
\begin{array}{cccc}
\text{CHO} & \left[\begin{array}{c}\text{CHOH}\end{array}\right. & \text{CHO} & \text{CH}_2\text{OH} \\
| & \| & | & | \\
\text{H–C–OH} & \text{COH} & \text{HO–C–H} & \text{CO} \\
| \quad \rightarrow & | \quad \rightarrow & | \quad \rightarrow & | \\
\text{H–C–OH} & \left.\text{HO–C–H}\right] & \text{HO–C–H} & \text{HO–C–H} \\
\vdots & \vdots & \vdots & \vdots \\
\text{D(+)glucose} & & \text{D(+)mannose} & \text{D(−)fructose}
\end{array}
$$

London Equation for Superconductors

The London equation replaces **Ohm's Law** in superconductors and is

$$j = - \frac{ne^2}{mc} A$$

where **A**, the vector potential, is defined by curl **A** =**H**; **H** is the magnetic field, n, e and m are the electronic density, charge and mass respectively and **j** is the current.

Sometimes the equation

$$\frac{d\mathbf{j}}{dt} = \frac{ne^2}{m} \mathbf{E},$$

where **E** is the electric field, is also called a London Equation. (F. London and H. London, *Proc. Roy. Soc. (London)* **A149**, 72 (1935).)

London Penetration Depth

The depth of penetration of an external magnetic field into a superconductor. It is given by

$$\lambda_L = \left(\frac{mc^2}{4\pi ne^2} \right)^{\frac{1}{2}}$$

where n, m and e are the electron density, mass and charge. If the thickness of a superconducting film is much less than λ_L then it will not exhibit the **Meissner Effect.**

London's Equation (for Intermolecular Attraction)

All molecules possess energy even at $0°K$ (null point energy) and this fact can only be explained by assuming that, even at this temperature, the nuclei and electrons vibrate in some way with respect to each other. As an instantaneous picture, the molecules will show various arrangements of the nuclei and electrons, these arrangements giving rise to dipole moments. A summation of these dipole moments over all the molecules will give a zero resultant. The cohesive force between molecules was attributed by London to the transient dipoles induced in molecules in phase with themselves by these temporary dipoles. The interaction energy has been calculated to a first approximation as

$$\Phi_D = -\frac{3}{4} \cdot \frac{h\nu_0 a^2}{r^6}$$

where ν_0 is the characteristic frequency of the molecules, a is equal to e^2/k where e is the electronic charge, k is the restoring force constant, and r is the equilibrium distance between the positive ends of the dipoles.

Lorentz *See* **Appendix**

Lorentz Force

The force **F** acting on a moving charge q in magnetic and electric fields is given by

$$\mathbf{F} = q \left(\mathbf{E} + \frac{\mathbf{v} \wedge \mathbf{B}}{c} \right)$$

where **v** is the velocity of the charge, **E** the electric field and **B** the magnetic induction. Often only the magnetic term in the force is referred to as the Lorentz force. The force can then be expressed in the non-vectorial form

$$F = \frac{q}{c} \, v B \sin \theta.$$

θ is the angle between the direction of motion of the charge and the direction of the magnetic field, and the force is perpendicular to both the direction of motion and the field.

Lorentz Gauge

For an electromagnetic field, knowledge of the electric and magnetic fields is not sufficient to determine uniquely the scalar potential ϕ and the vector potential $\mathbf{A}$ of the field. Hence an auxiliary condition is chosen (the gauge) giving the equations a symmetrical form. The Lorentz gauge is the condition

$$\nabla \mathbf{A} + \frac{1}{c} \frac{\partial \phi}{\partial t} = 0 \text{ (Gaussian units)}$$

$$\text{or } \nabla \mathbf{A} + \frac{1}{c^2} \frac{\partial \phi}{\partial t} = 0 \text{ (M.k.s. or S.I. units)}$$

Lorentz-Lorenz Law

$$\frac{n^2 - 1}{n^2 + 2} = \frac{4\pi}{3} \Sigma a$$

where n is the refractive index and Σa is the sum of polarizabilities of all particles present in unit volume. a is defined by

$$p = aE$$

where p is the electric dipole induced by an electric field E.

Lorentz Polarization Factor

In the scattering of x-rays by the atoms of a single crystal

$$LP = \frac{1 + \cos^2 2\theta}{2 \sin 2\theta}$$

where θ is the Bragg angle and LP is the Lorentz polarization factor. Tables of this factor versus θ are available.

A similar Lorentz polarization factor is available tabulated as a function of θ for scattering from a powder sample as in the **Debye-Scherrer Method**. Here

$$LP = \frac{1 + \cos^2 2\theta}{\sin^2 \theta \cos \theta}.$$

203

Lorentz's Theory of Light Sources

When Maxwell showed that light was an electromagnetic wave it was necessary to identify some electrical oscillator of atomic proportions to account for these waves.

Lorentz considered that electrons which are normally held in an equilibrium position are capable of vibration when displaced under the influence of a restoring force. The bound electron was supposed to be the source of the electromagnetic waves of light when executing damped vibrations. It was also capable of absorbing light when the frequency of the vibration and of the light were in agreement.

Lorentz Transformation

The relation connecting the distance and time intervals between two Lorentz frames. In these, the coordinates of an event are x, y, z, t and x', y', z', t' and the transformation gives

$$x^2 + y^2 + z^2 - c^2 t^2 = x'^2 + y'^2 + z'^2 - c^2 t'^2.$$

The transformation can be represented by a 4 by 4 matrix and all transformations of this type make up what is called a general Lorentz group.

Lorenz Number

From **Wiedemann Franz** and **Lorenz's Law**, the Lorenz number, L, is given by

$$L = \frac{\lambda}{\sigma T}$$

where λ is the thermal conductivity and σ the electrical conductivity. For metals the number is $k^2 \pi^2 / 3 e^2$ but in semiconductors it depends on the **Fermi Level** and on the type of carrier present.

Loschmidt Number

The number of molecules in 1 c.c. of a gas at $0^{\circ}C$ and 1 atmosphere pressure is known as the Loschmidt number. This quantity is equal to

$$L = \frac{N_A}{2 \cdot 24 \times 10^4} \simeq 2 \cdot 7 \times 10^{19}$$

where N_A is the Avogadro number.

Lossen Rearrangement

This rearrangement is related to the **Hofmann** and **Curtius Rearrangements**. Hydroxamine acids will rearrange to give isocyanates.

$$R-C-OH \rightarrow RNCO + H_2O$$
$$\parallel$$
$$NOH$$

Love Waves *See* Rayleigh Waves

Lüder's Bands

In the plastic flow of crystals, surface markings which appear during yield failure. These bands appear at obvious places of stress concentration, such as the shoulder of a specimen, when the upper yield point is reached. The boundaries of these surfaces divide the over-strained parts of the material from the parts which have not yielded.

Ludwig-Soret Effect Also called the **Soret Effect**

Lunge Scale *See* Baumé Scale

Lyddane-Sachs-Teller Relation

In an infinite ionic crystal, electromagnetic waves do not propagate in a frequency range $\omega_T < \omega < \omega_L$ where ω the frequency of the electric field. ω_L is also the longitudinal optical phonon frequency for the phonon wave vector **k** tending to zero and ω_T the transverse optical phonon frequency for a large wave vector. Then, if $\epsilon_r(0)$ is the static dielectric constant of the crystal, and $\epsilon_r(\infty)$ the dielectric constant at a frequency at which electronic polarizibility is effective but not ionic polarizibility, the Lyddane-Sachs-Teller Relation is

$$\frac{\epsilon_r(0)}{\epsilon(\infty)} = \frac{\omega_L{}^2}{\omega_T{}^2}.$$

Lyman Series *See* Balmer Series

M

MacCullagh's Formula

If X is the origin at a point of a distribution of mass or charge and P is an external point distance r from X, then the potential at P due to the distribution is given by

$$\Phi = \gamma \frac{M}{r} + \frac{\gamma}{2r^3} (A + B + C - 3I) + O\left(\frac{1}{r^4}\right)$$

where γ is the gravitational or electrostatic constant. M is the total mass or charge, I is the moment of inertia about XP and A, B and C are the principal moments of inertia about X.

Mach Line or Surface

The line or surface dividing the regions of subsonic and supersonic flow is called the Mach line.

Mach Number

The ratio of a velocity v to that of the velocity of sound, v_s, is termed the Mach number, M.

If $M = v/v_s < 1$ then v is said to be subsonic; if $v/v_s > 1$, v is supersonic.

Maclaurin-Cauchy Test

If $f(x) > 0$ and $f(x) \to 0$ steadily as $x \to \infty$, then $\int_1^\infty f(x)\, dx$ and $\sum_{n=1}^{\infty} f(x)$ converge or diverge together.

Maclaurin's Theorem

A function of x may be expanded as a polynomial in x and derivatives of $f(x)$ when $x = 0$.

$$f(x) = f(0) + f'(0)\frac{x}{1!} + f''(0)\frac{x^2}{2!} + \ldots + f^n(0)\frac{x^n}{n!} + R_n$$

$$R_n = f^{n+1}(\theta x)\frac{x^{n+1}}{(n+1)!}(1-\theta)^n \quad (0 < \theta < 1)$$

This is a special case of **Taylor's Theorem** for $h = 0$.

Madelung Constant

If ionic crystals are considered as composed of positively and negatively charged ions, the separation of pairs of ions being ρ_{ij}, Madelung's constant for a particular structure can be defined by

$$\alpha = \sum_{ij} (\pm)\rho_{ij}^{-1}$$

where the plus sign is used for unlike charges and the minus sign for like charges. It is used to calculate the binding energy of the lattice.

Maggi-Righi-Leduc Effect
Change of thermal conductivity of a conductor in a magnetic field.

Magnus' Effect
When a rapidly spinning body is hung up in an air current moving at right angles to its axis it is deflected at right angles to both the current and its axis and moves to the side where its peripheral motion is in the same direction as the current.

Majorana Effect
When light is passed through certain liquid or crystalline materials in a direction perpendicular to an applied magnetic field, the material becomes doubly refracting.

Majorana Force
An exchange force between nucleons in which charge and spin are exchanged (equivalent to exchange of position). If the nucleons are represented by wave-functions, then if the wave-functions are symmetric with respect to interchange of the space coordinate, the Majorana force is attractive; if the wave functions are antisymmetric, the Majorana force is repulsive. (*See also* **Bartlett, Heisenberg** *and* **Wigner Forces**.)

Makarov-Zemlianski-Prokin Method
Benzylidene chloride usually contains benzyl chloride and benzo-trichloride and consequently the hydrolysis product is contaminated with benzyl alcohol and benzoic acid. If, however, the hydrolysis is carried out by means of boric acid, only benzylidene chloride is hydrolysed, the other two being unaffected.

Malter Effect
The occurrence of large values of the coefficient of secondary electron emission in metals possessing a non-conducting surface film; notably in aluminium whose surface has been oxidized and then coated with caesium oxide. The surface film has a high resistance and becomes

positively charged, and there is field emission of electrons from the metal.

Malus' Law
If a plane polarized beam of light is allowed to fall on a polarizer, the intensity of the transmitted beam is proportional to the square of the cosine of the angle between the plane of polarization of the incident light and the plane of polarization that would be required for total transmission of the beam.

Malus' Theorem
A system of rays normal to a wave front remains normal to the wave front after any number of refractions and reflections.

Mandel'shtam Effect
An alternative description of **Brillouin Scattering.**

Mannich Reaction
Aldehydes and ketones that have an a hydroxyl react with formaldehyde and a secondary amine under weakly acidic conditions to give a dialkyl aminomethyl derivative of a carbonyl compound.

$$H_2C=O + HNR_2 \rightarrow H_2C.NR_2 \xrightarrow[H_2O]{H^+} H_2C=NR_2$$

$$\underset{OH}{|}$$

$$\downarrow RCOCHR'$$

$$RCOCH.CH_2-NR_2$$

$$\underset{R'}{|}$$

Mannich type reactions are general for other compounds which lose a proton such as acetylenes, nitroalkyls, phenols, thiophenes and pyrroles.

Markoff Process
A probability process in which development of the process depends entirely on its state at any one time and not on how that state arose. Every purely random process is a Markoff process.

Markownikoff's Replacement Rule
In the case of an unsymmetrical olefine, hydrogen halides can be added in two ways. Markownikoff's Rule states that the halogen is affixed to the carbon atom carrying the smaller number of hydrogen

atoms. It has been shown, however, that normal addition can frequently be reversed in the presence of peroxy compounds.

$$CH_2=CHBr + HBr \xrightarrow{\text{Peroxide Catalyst}} \begin{array}{c} CH_3.CHBr_2 \\ \\ BrCH_2.CH_2Br \end{array}$$

Marriotte's Law (Boyle's Law) *See* **Boyle's Law**

Marsh Test for Arsenic
The substance to be tested is treated with arsenic-free zinc and hydrochloric acid to convert any arsenic present to arsine. The arsine is dried with $CaCl_2$ and freed from H_2S by lead acetate and is then decomposed by heat to give a characteristic mirror. Antimony gives a similar stain, which may be distinguished from the arsenic stain because it is not soluble in sodium hypochlorite solution.

Marx Effect
The reduction in the energy of photoemission by the simultaneous incidence of radiation of lower frequency than that producing the emission.

Mascheroni's Constant *See* **Euler's Constant**

Mason's Theorem
If S is a convergent operator, the equation

$$f = g + S.f$$

admits of a continuous solution, given by

$$f = g + S.g + S^2.g + S^3.g + \ldots$$

Let ϕ be any function continuous in a region R, then S is a convergent operator if
   (a) $S\phi$ is a continuous function in R;
   (b) $\phi + S\phi + S^2\phi + \ldots$ converges and is a continuous function in R;
   (c) the result of the operation S on the function represented by the series (b) is equal to the result of the term-by-term operation on the series.

Massieu Function

A thermodynamic function J is given by

$$J = -(U/T) + S$$

where U is internal energy and S entropy. It is equal to minus the quotient of the **Helmholtz Function** and the temperature T.

Mathieu's Function

Mathieu's functions are solutions of the Mathieu equation;

$$\frac{d^2\chi}{d\phi^2} + (b - h^2 \cos^2 \phi) \chi = 0$$

which occurs in problems involving systems with elliptical symmetry.

Depending on the value of b, the functions are periodic, stable or unstable.

Matteuci Effect

The appearance of an electric potential difference between the ends of a ferromagnetic specimen when twisted in a magnetic field. (*See also* **Barnett Effect.**)

Matthias' Rules

Empirical rules for the transition temperatures of superconducting metals and alloys:

I The transition temperature of an element depends regularly on its position in the periodic table but the transition metals show entirely different dependence on their valency to that exhibited by metals with completed d-shells.

II For alloys between transition metals, or between non-transition metals, the transition temperature of the alloys is a smooth function of the mean valency v, weighted according to the atomic fractions of the constituent metals.

III For elements and alloys with closed d-shells, the transition temperature rises from $0°K$ at $v = 2$ to a maximum at $v = 5$.

IV In the transition series, the transition temperature has a maximum at $v = 5$ and $v = 7$ with a sharp minimum at $v = 6$.

Matthiessen's Rule

If the concentration of impurity atoms in a metal is small, the resistivity caused by scattering of electron waves by impurity atoms is independent of temperature. This resistivity is additive to that caused by the thermal motion of the periodic lattice of the metal, this latter resistivity being a function of temperature.

Matthiessen's Standard
A measure of the conductivity of copper expressed as a percentage of 0·1539 (if hard drawn) or 0·1508 (if annealed) ohms for a specimen of length 1 metre and of weight 1 gramme.

Maupertuis' Principle
The motion of a mechanical system is entirely determined by the principle of least action in which the integral $S = \int_{t_1}^{t_2} L(q_r, \dot{q}_r, t)\mathrm{d}t$ must be a minimum.
$L = L(q_r, \dot{q}_r, t)$ is the **Lagrangian** and q_r is a generalized coordinate.
The principle is analogous to **Fermat's Principle** in optics.

Maxwell *See* Appendix

Maxwell-Boltzmann Distribution Law
The number of particles of a non-quantized system of which the velocity lies between v and $v + \mathrm{d}v$ is

$$\mathrm{d}N = 4\pi N \left(\frac{m}{2\pi kT}\right)^{3/2} e^{-mv^2/2kT}\,\mathrm{d}v$$

where N = total number of particles,
v = random velocity,
m = mass of particle.

Maxwell-Boltzmann Statistics
A method of studying the probabilities of the states of a system of non-quantized particles, defined by the average values of the position, velocity or energy coordinates, in a small but finite volume of the system.

Maxwell Effect (Flow Birefringence)
If a liquid is viscous, consists of anisotropic molecules, and flows in such a way that a shearing velocity gradient occurs, then it exhibits optical birefringence. If the liquid has velocity v_x (i.e. velocity in the

x direction only), then for light travelling in the z direction the principal planes of polarization are at $45°$ to the x and y axes, except for very high velocity gradients. The difference in refractive index is given by

$$\Delta n = c\left(\frac{\partial v_x}{\partial y}\right)$$

where c is **Maxwell's constant**, which is a measure of birefringence.

Maxwell Equation (for Refractive Index)

For a non-absorbing non-magnetic material, the dielectric constant ϵ_r and refractive index n are related by the equation

$$\epsilon_r = n^2.$$

Maxwell Primaries

The magenta, green and cyan colours forming the primary colours in Maxwell's three-colour additive synthesis.

Maxwell's Demons

Spontaneous changes in temperature and pressure such as to invalidate the Second Law of Thermodynamics could possibly be brought about by the intervention of intelligent beings: 'Maxwell's demons'. Thus by means of a one-way valve they might collect more molecules in one part of a system than in another and thereby cause an increase of pressure. In fact experience shows that such 'demons' do not exist in closed systems in temperature equilibrium.

Maxwell's Equations

A set of four equations of classical electromagnetic theory. They relate certain vector quantities in a space which experiences changes in electric and magnetic fields. If **H** is the magnetic field, **B** the magnetic induction, **E** the electric field, **D** the electric displacement,

ρ the electric charge density and $\mathbf{j}$ the conduction current density, then Maxwell's equations are;

Gaussian units	*M.k.s. or S.I. units*
$\text{curl } \mathbf{H} = \dfrac{1}{c}\dfrac{\partial \mathbf{D}}{\partial t} + \dfrac{4\pi}{c}\mathbf{j}$	$\text{curl } \mathbf{H} = \dfrac{\partial \mathbf{D}}{\partial t} + \mathbf{j}$
$\text{curl } \mathbf{E} = -\dfrac{1}{c}\dfrac{\partial \mathbf{B}}{\partial t}$	$\text{curl } \mathbf{E} = -\dfrac{\partial \mathbf{B}}{\partial t}$
$\text{div } \mathbf{B} = 0$	$\text{div } \mathbf{B} = 0$
$\text{div } \mathbf{D} = 4\pi\rho$	$\text{div } \mathbf{D} = \rho$

Maxwell's Rule
Every part of an electric circuit is acted upon by a force which tends to move it in such a direction as to enclose the maximum amount of magnetic flux.

Maxwell's Theorem
In an elastic structure, if a load applied at one point produces a given deflection at another point then the same load applied at the point where the deflection was first measured will produce the same deflection as before at the point where the load was first applied.

Maxwell's Theory of Light
Clerk Maxwell showed in 1860 that the propagation of light could be regarded as an electromagnetic phenomenon, the wave consisting of an advance of coupled electric and magnetic forces. If an electric field is varied periodically, a periodically varying magnetic field is obtained which in turn generates a varying electrical field and thus the disturbance is passed on in the form of a wave.

Maxwell's Thermodynamic Equations
For a thermodynamic system,

$$\left(\frac{\partial p}{\partial T}\right)_V = \left(\frac{\partial S}{\partial V}\right)_T \qquad\qquad \left(\frac{\partial V}{\partial T}\right)_V = -\left(\frac{\partial S}{\partial p}\right)_T$$

$$\left(\frac{\partial p}{\partial S}\right)_V = -\left(\frac{\partial T}{\partial V}\right)_S \qquad\qquad \left(\frac{\partial V}{\partial S}\right)_p = \left(\frac{\partial T}{\partial p}\right)_S$$

where p is the pressure, V the volume, T the absolute temperature and S the entropy.

Maxwell Stress Tensor

Maxwell considered the force between electric charges to be transmitted in a continuous way by means of tubes of force which are considered as being in a state of tension. Hence, if a system in electrostatic equilibrium is divided into two parts by surface S, the force exerted on one part by the other must pass through the surface S. Maxwell showed this force to be given by

$$F = \iint_S T_n \, dS$$

where T_n is the component matrix of the Maxwell stress tensor.

$$T \equiv \frac{1}{4\pi} \begin{bmatrix} \epsilon_r E_x{}^2 - \dfrac{\mathbf{E}^2}{2}\left(\epsilon_r - \dfrac{d\epsilon_r}{d\sigma}\sigma\right) & \epsilon_r E_x E_y & \epsilon_r E_x E_z \\ \epsilon_r E_x E_y & \epsilon_r E_y{}^2 - \dfrac{\mathbf{E}^2}{2}\left(\epsilon_r - \dfrac{d\epsilon_r}{d\sigma}\sigma\right) & \epsilon_r E_y E_z \\ \epsilon_r E_x E_z & \epsilon_r E_y E_z & \epsilon_r E_z{}^2 - \dfrac{\mathbf{E}^2}{2}\left(\epsilon_r - \dfrac{d\epsilon_r}{d\sigma}\sigma\right) \end{bmatrix}$$

where $\mathbf{E}$ is the electric field strength and E_x, E_y and E_z its x, y and z components, ϵ_r the dielectric constant and σ the charge density.

Meerwein-Ponndorf-Verley Reduction

When a carbonyl compound is heated with aluminium propoxide in isopropanol, the propoxide is oxidized to acetone and the carbonyl group is reduced to the alcohol.

$$R_2CO + (CH_3)_2CHO\,Al_{1/3} \rightleftharpoons (R_2CHO)Al_{1/3} + CH_3.CO.CH_3$$

$$\downarrow \substack{\text{dilute} \\ H_2SO_4}$$

$$R_2CHOH$$

The acetone is removed from the equilibrium mixture by slow distillation. This reaction is specific for the carbonyl group and may be used in compounds containing other reducible substituents.

Meerwein Reaction

Diazonium halides in the presence of cuprous chloride add to a carbon-carbon double bond. The reaction works well for α, β

unsaturated carbonyl compounds and nitriles. It is usually carried out in acetone solution with catalytic quantities of cuprous chloride (e.g. **Sandmeyer Reaction**).

Mehler's Integrals

$$P_m(\cos\theta) = \frac{2}{\pi} \int_0^\theta \frac{\cos(m+\frac{1}{2})\phi}{\sqrt{\{2(\cos\phi - \cos\theta)\}}} d\phi \quad 0 < \theta < \pi$$

for all values of m. If m is a positive integer

$$P_m(\cos\theta) = \frac{2}{\pi} \int_\theta^\pi \frac{\sin(m+\frac{1}{2})\phi}{\sqrt{\{2(\cos\theta - \cos\phi)\}}} d\phi$$

where P_m is the **Legendre Coefficient**.

Meissner Effect

A bulk superconductor in a weak magnetic field exhibits perfect diamagnetism. Thus, if a specimen in the normal state is cooled through the transition temperature for superconductivity, the magnetic flux originally present in the sample is expelled, except for slight penetration at the surface. See also **London Equation for Superconductivity** and **London Penetration Depth**.

Mellin Transforms

If $f(x)$ is defined for $(0 \leqslant x \leqslant \infty)$ if

$$\int_0^\infty |f(x)|^2 \, x^{-2s-1} \, dx$$

converges for $s > s_0$, and if $M(p) = \displaystyle\int_0^\infty f(x) \, x^{p-1} \, dx$

then $M(p)$ is called the Mellin transform of $f(x)$ and conversely

$$f(x) = \frac{1}{2\pi j} \int_{-j\infty+s}^{j\infty+s} M(p) \, x^{-p} \, dp \quad (s > s_0; x > 0)$$

215

Mendeleef's Periodic Table

In 1869, three years after the publication of **Newland's Law of Octaves**, Mendeleef presented a paper to the Russian Chemical Society 'On the correlation of the Properties and Atomic Weights of the Elements'.

His 'Table of Elements' arranged the then known elements into five periods leaving blank spaces where necessary to establish periodicity and predicting that these blanks would be found to correspond with elements as yet unknown.

Mendius Reaction

Alkyl cyanides are reduced by nascent hydrogen from sodium and ethyl alcohol to give primary amines.

$$R.CN + 4H \rightarrow R.CH_2.NH_2$$

Menelaus' Theorem

If ABC is a triangle and PQR a straight line cutting BC produced, CA, and also AB, at P, Q and R respectively, then

$$\frac{BP}{CP} \cdot \frac{CQ}{QA} \cdot \frac{AR}{RB} = 1$$

Mercator's Projection

A projection in which a sphere's surface (e.g. the earth's surface) is projected on to a cylinder taken about the equator. This transverse Mercator projection becomes increasingly inaccurate with increasing latitude, but alternative projections (oblique Mercator's projections) are made, where the cylinder's axis is taken at an angle other than $90°$ to the equator.

Mersenne Numbers

Numbers M given by $M = 2^p - 1$, where p is a prime number.

Mersenne's Law

When a string is set into vibration, the frequency of the fundamental varies directly as the square root of the tension and inversely both as the length and as the square root of the mass per unit length of the string.

Mertens' Theorem

If Σu_n and Σv_n are convergent with sums s and t respectively, and one of these series, say Σu_n, is absolutely convergent, then Σd_n is convergent and its sum is st, where

$$\Sigma d_n = u_1 v_n + u_2 v_{n-1} + \ldots + u_n v_1$$

Meusnier's Theorem
The osculating plane of a curve C at a point P_1 on a surface is the plane containing the limiting circle through P_1 and through two distinct points P_2 and P_3 (with P_2 and P_3 approaching P_1) and containing the tangent to C at P_1. Then Meusnier's theorem states that at P_1 the curvature κ of C is given by

$$\kappa = \left| \frac{\kappa_N}{\cos a} \right|$$

where a is the angle between the osculating plane and the normal section plane through the tangent, and κ_N is the normal section curvature.

Meyer Atomic Volume Curve
If the atomic volumes of the elements are plotted against their atomic numbers a periodicity is revealed. The alkali metals are seen to be on the peaks whilst the transition elements appear in the valleys.

Michaelis Reaction
Phosphonic esters can be prepared from the reaction of a sodium salt of an alkyl phosphonate with an alkyl halide. (See also **Arbusov Reaction**.)

$$(RO)_2 \, P - ONa + R'X \rightarrow (RO)_2 \, P = O + NaX$$
$$|$$
$$R'$$

Michael Reaction
This reaction is a generalization of the **Claisen Condensation**. Other active methylene compounds such as ethyl phenyl acetate, benzyl cyanide and ethyl cyanoacetate also add to $a\beta$ unsaturated ketones, esters, nitriles and sulphones.

Mie-Grüneisen Equation of State
An equation of state with particular application at high pressures.

$$p - p_0 = \frac{\gamma}{V} \, (U - U_0)$$

where p_0 and U_0 are the pressure and internal energy at $0°K$ and are

217

functions of the volume at $0°K$ and γ is the Grüneisen Number which is a function of volume V only.

Miller-Bravais Indices

The **Miller Index** notation has been extended to the hexagonal system where directions x, y and u refer to three horizontal axes at an angle of $120°$ to each other and each normal to the vertical z axis. Then, the Miller-Bravais indices, represented by symbols $h\ k\ i\ l$, refer to the reciprocals of intercepts on these axes. The first three indices are not independent as $h + k + i = 0$. These symbols are therefore sometimes written as $h\ k \cdot l$ or $h\ k * l$.

Miller Effect

In a single electronic-tube voltage amplifier, the apparent capacitance between the grid and anode is multiplied because of feedback by $(1 - A)$ where A is the complex gain of the stage. A is usually large and negative in sign, a positive voltage at the grid giving rise to a negative voltage at the anode. The effect also occurs between the collector and the base of a transistor amplifier.

Miller Index

In order to represent any particular crystal plane in terms of the crystallographic axes and the intercepts of the plane on these axes, the reciprocal of the intercepts are taken on the three axes in terms of

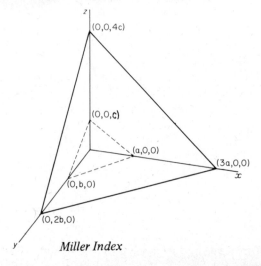

Miller Index

218

the lattice constants, and these reciprocals are reduced to the smallest three integers having the same ratio. These integers are the Miller indices of the plane. Thus, for the example, the plane shown in the diagram has intercepts 3a, 2b, 4c, whose ratios in terms of the lattice constants are 3, 2, 4, and whose reciprocals are $\frac{1}{3}$, $\frac{1}{2}$, $\frac{1}{4}$. The Miller indices are therefore 4, 6, 3.

Miller's Law

In a crystal, the sine ratio of four faces in a zone is a commensurable number. (A zone consists of a set of planes whose intersecting edges are parallel.) If P_1, P_2, P_3 and P_4 are the four faces, and θ_{12} is the angle between P_1 and P_2 etc., then the sine ratio is given by

$$\frac{\dfrac{\sin \theta_{12}}{\sin \theta_{13}}}{\dfrac{\sin \theta_{42}}{\sin \theta_{43}}} = \frac{\dfrac{U_{12}}{U_{13}}}{\dfrac{U_{42}}{U_{43}}} = \frac{\dfrac{V_{12}}{V_{13}}}{\dfrac{V_{42}}{V_{43}}} = \frac{\dfrac{W_{12}}{W_{13}}}{\dfrac{W_{42}}{W_{43}}}$$

where $U_{12} = (k_1 l_2 - l_1 k_2)$, $V_{12} = (l_1 h_2 - h_1 l_2)$ and $W_{12} = (h_1 k_2 - k_1 h_2)$ etc., hkl being the **Miller Indices** of the faces.

Millington Reverberation Formula

In acoustics, a formula giving the reverberation time T in seconds as

$$0{\cdot}05 \frac{V}{\sum\limits_{r=1}^{n} S_r \ln (1 - a_r)}$$

where
$\qquad V = $ Volume in cubic feet
$\qquad a_1, a_2, a_3, \ldots = $ absorption coefficients of areas $S_1, S_2, S_3 \ldots$

Millon's Base

Dihydroxomercuric-ammonium hydroxide

$$\begin{bmatrix} HO{-}Hg \\ HO{-}Hg \end{bmatrix} {>} NH_2 \end{bmatrix} OH$$

produced by the action of ammonia solution on yellow mercury oxide.

219

Millon's Reaction
When a protein is heated with a solution of mercuric nitrite and nitrate in nitrous and nitric acids (Millon's Reagent) a red coloration or precipitate is formed.

Minkowski's Inequality

$$\left[\int_a^b |f_1 + f_2|^p \, dx\right]^{1/p} \leqslant \left[\int_a^b |f_1|^p \, dx\right]^{1/p} + \left[\int_a^b |f_2|^p \, dx\right]^{1/p}$$

$$(1 \leqslant p < \infty)$$

The inequality implies a similar inequality for sums of convergent series;

$$\left[\sum_j |a_j + b_j|^p\right]^{1/p} \leqslant \left[\sum_j |a_j|^p\right]^{1/p} + \left[\sum_j |b_j|^p\right]^{1/p}$$

See also **Hölder's Inequality**.

Minkowski Space
Four-dimensional space involving the x, y and z dimensions of real space and the dimension of time. The **Lorentz Transformation** can be considered as a rotation in Minkowski space.

Misterlich Law of Isomorphism
Substances which are similar in crystalline form and in chemical properties can usually be represented by similar formulae.

Möbius Band
A one-sided surface formed by rotating one end of a strip of paper through $180°$ and then fastening the ends together.

Mohr's Litre
The volume occupied by 1 kg of distilled water when weighed in air against brass weights at $17\cdot5°$C. It is approximately equal to $1\cdot002$ true litres.

Mohs' Hardness Scale
A scale of hardness of materials based on a scratching test wherein any material can be scratched by the material which follows it in the scale.

Molisch's Test for Carbohydrates
A sugar when mixed with a-naphthol and treated with concentrated sulphuric acid gives a violet coloration

Møller Scattering
Electron scattering by another electron.

Mond Process for Nickel
After treatment, the nickel-bearing pyrites are reduced to a matte containing 50% Ni, 30% copper and 20% sulphur. This matte is roasted to yield the oxides which are extracted with sulphuric acid to give copper sulphate which is recrystallised and sold whereas the nickel oxide is hardly attacked at all. The residue of nickel oxide is reduced by water gas. The metal so formed still contains some copper and is refined by passing over it carbon monoxide gas at 60°C, when nickel carbonyl $Ni(CO)_4$, which is volatile, is formed. The vapour is passed over nickel shot heated to 180°C where it decomposes to give pure nickel and carbon monoxide which is used again.

Monge's Theorem
By definition, centres of similitude are the points of intersection of lines joining the ends of parallel radii of two coplanar circles. Monge's theorem then states that the three outer centres of similitude of three circles taken in pairs are collinear. Any two of the inner centres of similitude are collinear with one of the outer centres.

Morse Equation
The potential energy of a diatomic molecule (Φ_r) according to P. M. Morse (1929) may be approximated by

$$\Phi_r = D\left(1 - \exp[-a(r - r_e)^2]\right)$$

where D is the dissociation energy, r is the distance apart of the atoms and r_e is the distance at equilibrium. a is a constant given by

$$a = \left(\frac{8\pi^2 \mu x \omega e c}{h}\right)^{\frac{1}{2}}$$

where μ is the reduced mass, ω the equilibrium angular frequency and x the anharmonicity constant.

221

Moseley's Law

If x-rays are allowed to fall upon a substance a series of characteristic x-ray lines are produced. Moseley showed that the frequency v of the line belonging to any particular x-ray series is quantitatively related to the atomic number Z by the relationship

$$\sqrt{v} = a(Z - \sigma)$$

where a is the proportionality constant and σ is the same for all lines of a given series.

Mossbauer Effect

The Mossbauer effect is the recoiless emission and resonant absorption of gamma radiation. When a nucleus decays from an excited state by emitting a gamma ray, the energy of the gamma ray is given by the energy level of the excited state less the energy imparted to the nucleus on recoil. The effect of the recoil of the nucleus is to reduce the gamma ray energy below the inter-level energy such that a second nucleus in a lower state cannot be excited by gamma ray absorption. However, in 1958, Mossbauer found that in certain systems (e.g. containing Fe^{57}), the recoil momentum is taken up by the whole crystal in which the nucleus is embedded. In this case, because of the much larger recoil mass, the energy of the emitted gamma ray is sufficient for it to excite a further nucleus. The uncertainty spread in the frequency of the gamma ray is then as small as 1 part in 10^{13}. Hence the resonant absorption condition will be very sensitive to any relative velocity imparted between the source and absorber, and a relative velocity of 1–10 mm per second can throw the absorber out of resonance. Thus using separate source and absorber, the effect can be used for energy resolution when, for instance, the energy levels of the source are split by the **Zeeman Effect**, and can also be used to repeat the Michelson-Morley experiment.

Mott Exciton *See* Frenkel Exciton

N

Naperian Logarithm

If $y = e^x$, where $e = \lim_{n \to \infty} [1 + (1/n)]^n = 2 \cdot 71828182 \ldots$, then x is termed the Naperian logarithm of y.

Napier's Rules

In the solution of right-angled spherical triangles: If a, b, c are the sides of the triangle and a, β, γ the angles opposite a, b, c respectively, c being the side opposite γ the right angle, then the circular parts of the triangle are a, b, $\pi/2 - c$, $\pi/2 - a$, $\pi/2 - \beta$. If these parts are supposed ranged inside a circle in the order in which they naturally occur with respect to the triangle, any one of these parts may be selected and called the middle part; the two parts next to it are the adjacent parts and the two remaining parts are called the opposite parts. Napier's rules are:

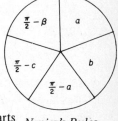

Napier's Rules

The sine of the middle part is equal to the product of the tangents of the adjacent parts.

The sine of the middle part is equal to the product of the cosines of opposite parts. *See also* **Delambre's Analogies**.

Navier-Stokes Equation

Equation of motion for a viscous fluid. If the fluid is incompressible;

$$\frac{\partial \mathbf{v}}{\partial t} + \mathbf{v} \nabla \mathbf{v} = \mathbf{F} - \frac{1}{\rho} \operatorname{grad} p + \eta \nabla^2 \mathbf{v}$$

where $\mathbf{v}, p, \mathbf{F}$ and ρ are as defined for **Euler's Equation** and η is the dynamic viscosity. For a compressible fluid, three equations are required of the type;

$$\frac{\partial v_x}{\partial t} = F_x - \frac{1}{\rho}\frac{\partial p}{\partial x} + \frac{1}{\rho}\frac{\partial}{\partial x}\left[\eta\left(2\frac{\partial v_x}{\partial x} - \frac{2}{3}\nabla\mathbf{v}\right)\right] + \frac{1}{\rho}\frac{\partial}{\partial y}\left[\eta\left(\frac{\partial v_x}{\partial y} + \frac{\partial v_y}{\partial x}\right)\right]$$

$$+ \frac{1}{\rho}\frac{\partial}{\partial z}\left[\eta\left(\frac{\partial v_z}{\partial x} + \frac{\partial v_x}{\partial z}\right)\right].$$

v_x, v_y and v_z are component velocities in the x, y and z directions in Cartesian coordinates and F_x is the component field of force in the x direction.

Néel Temperature
The temperature at which the spins of the neighbouring ions in an antiferromagnetic material cease to be ordered antiparallel as the temperature of the material is increased.

Nef Reaction
If a primary or secondary nitro-compound is converted to the aciform by means of an alkali and then hydrolysed with 25% sulphuric acid, aldehydes or ketones are produced with the evolution of nitrous oxide.

$$2RCH=NOONa + 2H_2SO_4 \rightarrow 2RCHO + N_2O + 2NaHSO_4 + H_2O$$

Neil's Parabola
A curve whose equation is

$$y = a\,x^{3/2}$$

Nelson-Riley Function
In the **Debye and Scherrer Method** of x-ray crystal analysis, the percentage error in the calculation of the lattice parameter for given error in the measurement of diffraction angle approaches zero for a **Bragg Angle** of $\theta = 90°$. By plotting the calculated lattice parameter against the function

$$\frac{1}{2}\left(\frac{\cos^2\theta}{\sin\theta} + \frac{\cos^2\theta}{\theta}\right)$$

a linear extrapolation can be made to $\theta = 90°$. This function (available tabulated) was obtained by Nelson and Riley and independently by Taylor and Sinclair.

Neper *See* **Appendix**

Nernst Approximation Formula
Nernst was able to deduce the equilibrium constant of a gaseous reaction thermodynamically. From this relation, by a series of approximations which are now known to be inaccurate, he was able to obtain an equation which, though crude, can give useful information when no other data are available. The equation is

$$\log K' = \frac{-\Delta H}{4\cdot58T} + 1\cdot75 \, \Sigma \, v_1 \log T + \Sigma \, v_1 \, C_1$$

where ΔH may be taken as the value at room temperature; $\Sigma \, v_1$ is taken for the products minus reacting substances. For heterogeneous reactions, $\Sigma \, v_1$ and $\Sigma \, v_1 \, C_1$ refer only to the gases, but ΔH to the whole reaction. C_1 is the conventional chemical constant, an empirical quantity, and v_1 is the number of reacting molecules of species 1.

Nernst Effect (or **Nernst-Ettingshausen Effect**)
If a temperature gradient is established in a conductor which is placed in a magnetic field that is orthogonal to the temperature gradient, then electrons tending to move from the hot to the cold end are initially deflected by the magnetic field. This produces a transverse electric field which prevents further deflection and which is given by

$$E_y = Q \, B_z \, \frac{dT}{dx}$$

where E_y is the Nernst electric field, Q is the Nernst coefficient, B_z the magnetic flux density and dT/dx the longitudinal temperature gradient.

The above effect is sometimes referred to as the transverse Nernst-Ettingshausen effect. The variation of the **Seebeck Effect** in a magnetic field is then called the longitudinal Nernst-Ettingshausen effect.

Nernst Equation for E.M.F.
The potential $E_\pm$ of any reversible electrode immersed in a solution of a single ion of valency $Z_\pm$ is given by the equation

$$E_\pm = E_0 \mp \frac{RT}{Z_\pm F} \ln C_\pm$$

where E_0 is the standard electrode potential, and $C_\pm$ is the concentration of the ions.

Nernst Heat Theorem
For a homogeneous system, the rate of change of free energy with temperature, and also the rate of change of heat content with temperature, approaches zero as the temperature approaches absolute zero.

Nernst-Lindemann Equation
An equation relating the specific heats C_p and C_V at constant pressure and volume respectively for a pure substance. It is used for calculating the variation of C_V with temperature.

$$C_p - C_V = A\, C_p^2\, T$$

where

$$A = \frac{V\,\gamma^2}{\kappa\, C_p^2}$$

γ is the volume expansivity and κ the compressibility. A is a constant which can be obtained from the values of V, γ, κ and C_p for a single value of T. The equation is also called **Lewis' Equation**.

Nernst's Distribution Law
The partition of a substance between two solvents is governed by a law

$$\frac{C_1}{C_2^{1/n}} = \text{const.}$$

where C_1, C_2 are the concentrations of the substance in solvents 1, 2 and where the substance is present in monomolecular form A in solvent 1 and as polymeric molecules A_n in solvent 2.

Nernst-Simon Statement (of the Third Law of Thermodynamics)
The entropy change associated with any isothermal reversible process
of a condensed system approaches zero as the temperature approaches
zero. See also **Nernst Heat Theorem.**

Nernst Solution Pressure
When an electrode is immersed in a solution of its ions a potential
difference exists between the solution and the electrode. The first
theoretical explanation of this was due to W. Nernst (1881) who
considered that all metals and hydrogen possess a property called
solution pressure by virtue of which they tend to pass into solution
as positive ions. In doing so they leave the metal negatively charged
and an electrical double layer is set up at the electrode. Due to the
charge the amount of metal which goes into solution is very small.
This tendency, it was postulated, was opposed by a tendency for
ions to leave the solution and become metal due to osmotic pressure.
From a consideration of an equilibrium between these two processes
Nernst was able to obtain, using the methods of thermodynamics,
an equation for the electrode potential (*see* **Nernst Equation**).

Nernst-Thomson Rule
J. J. Thomson and W. Nernst in 1893 pointed out that in a medium of
low dielectric constant the electrostatic attraction between the anions
and cations of a dissolved electrolyte will be large whilst in a solvent of
high dielectric constant the attraction will be small. Hence solvents of
high dielectric constant will favour dissociation whilst solvents of low
dielectric constant will have small dissociating influence.

Nernst Zero of Potential
An arbitrary zero of electrode potential defined as that potential
corresponding to the reversible equilibrium between hydrogen gas at
one standard atmosphere pressure and hydrogen ions at unit activity.

Nesmeyanov Reaction
When mercury chloride is added to a solution of a diazonium chloride a
complex addition product is precipitated. In the Nesmeyanov reaction
this complex is suspended in acetone and heated in the presence of
copper powder to give an aryl mercuric chloride.

$$ArN_2Cl.HgCl_2 + 2Cu \rightarrow ArHgCl + N_2 + 2CuCl$$

Neumann-Kopp Rule

The heat capacity of a gram-atom of a solid phase can be considered as the weighted sum of the heat capacities of the elements forming the phase. The rule is applicable to compounds formed from elements in the solid or liquid states.

Neumann Problem

In potential theory, the determination of a function, harmonic in a region, when the boundary values of its normal derivatives are given (*see* **Dirichlet Problem**).

Neumann's Bessel Functions of the Second Kind *See* Bessel Functions

Neumann's Expansion

Any polynomial in x can be expanded as a series of **Legendre's Coefficients**. If f(x) is of degree n,

$$f(x) = \sum_{k=0}^{k=n} a_k \, P_k(x)$$

and

$$a_k = (k + \tfrac{1}{2}) \int_{-1}^{+1} f(x) \, P_k(x) \, dx$$

Neumann's Formula

This gives the mutual inductance between two circuits in the form

$$M_{12} = \oint_1 \oint_2 \mu \, \frac{dl_1 \cdot dl_2}{r}$$

where dl_1 and dl_2 are vector elements of currents in the two circuits, r is the distance between them and μ is the permeability. The integration is taken over all possible combinations of the scalar product of dl_1 and dl_2.

Neumann's Law

Whenever the flux of magnetic induction enclosed by a circuit is changing, there is an e.m.f. acting round the circuit, in addition to the

228

e.m.f. of any batteries which may be in the circuit. The amount of this additional e.m.f. is equal to the rate of diminution of the flux of induction enclosed by the circuit.

Neumann's Method for Organic Phosphorus
To the sample add 1 drop of concentrated nitric acid and 12 drops of sulphuric acid. Heat strongly with constant agitation. The solution turns black and more nitric acid is added until oxidation ceases (no further evolution of N_2O_4). Cool. Add 5 c.c. of distilled water, 1 drop of methyl red and 0·880 ammonia dropwise until the solution becomes yellow. Add 2 c.c. of ammonium molybdate and boil. A yellow precipitate of ammonium phosphomolybdate indicates the presence of phosphorus.

Neumann's Principle
The symmetry elements of any physical property of a crystal must include the symmetry elements of the point group of the crystal.

Neumann's Series
An arbitrary function expanded in terms of **Bessel Functions**:

$$\phi(x) = \Sigma a_n \, J_n(x)$$

is called Neumann's Series.

Newland's Law of Octaves
If the elements were arranged in rising order of their atomic weights, starting from any one element, the eighth above it was of similar nature to the first. This rule, of course, breaks down after the first two short periods of the Periodic Table. Although only 62 elements were known in Newland's time (1863) he made no provision for the inclusion of new elements when enunciating this rule.

Newton *See* Appendix

Newton-Cotes Formula
A method of numerical integration in which the range of integration is divided into equally spaced intervals of width h. Then,

$$\int_{x_0}^{x_0+nh} y(x) \, dx \approx A_0 y_0 + A_1 y_1 + \ldots A_n y_n = A.$$

229

However, for $n > 6$, it is usual to add m sums of $n' \leqslant 6$:

$$\int_{x_0}^{x_0+mn'h} y(x)\,dx = \int_{x_0}^{x_0+n'h} y(x)\,dx + \int_{x_0+n'h}^{x_0+2n'h} y(x)\,dx + \dots$$

If $n' = 1$, this gives for each sub-interval

$$A_0 = A_1 = h/2,$$

so that when all m successive sub-intervals are included

$$A = h\left[\frac{y_0}{2} + y_1 + y_2 \dots y_{n-1} + \frac{y_n}{2}\right]$$

This is called the trapezoidal rule. $n' = 2$ and $n' = 6$ give, respectively, **Simpson's** and **Weddle's Rules.**

Newton-Gauss Interpolation Formula

A formula analogous to **Gregory's Interpolation Formula.**

$$f(x) = f(x_0) + \frac{x - x_0 \Delta f(x_{1/2})}{h} + \frac{(x - x_0)(x - x_0 - h)\Delta^2 f(x_0)}{2!\,h^2}$$

$$+ \frac{(x - x_0)(x - x_0 - h)(x - x_0 + h)\Delta^3 f(x_{1/2})}{3!\,h^3} + \dots$$

$$+ \frac{(x - x_0)(x - x_0 - h)(x - x_0 + h)\dots\{x - x_0 - (n-1)h\}\{x - x_0 + (n-1)h\}}{(2n-1)!\,h^{2n-1}}\Delta^{2n-1}f(x_{1/2})$$

$$+ \frac{(x - x_0)(x - x_0 - h)(x - x_0 + h)\dots\{x - x_0 - (n-1)h\}\{x - x_0 + (n-1)h\}\{x - x_0 - nh\}}{2n!\,h^{2n}}\Delta^{2n}f(x_0)$$

where

$$\Delta f(x_{1/2}) = f(x_0 + h) - f(x_0) \qquad \Delta f(x_{-1/2}) = f(x_0) - f(x_0 - h)$$
$$\Delta^2 f(x_0) = \Delta f(x_{1/2}) - \Delta f(x_{-1/2})$$
$$\Delta^3 f(x_{1/2}) = \Delta^2 f(x_0 + h) - \Delta^2 f(x_0)\ \text{etc.}$$

Newtonian and Non-Newtonian Liquids

Liquids which show a linear relationship between velocity gradient and shearing stress are termed Newtonian liquids. Complex substances and high molecular weight and colloidal solutions do not exhibit this simple behaviour; the velocity gradient increases superlinearly with shearing stress, i.e. the viscosity of the liquid decreases with shearing stress. Such liquids are called non-Newtonian.

Newton-Raphson Formula

If it is required to find the root x of the equation $f(x) = 0$, then if x_n is a given approximation, an improved approximation is given by

$$x_{n+1} = x_n - \frac{f(x_n)}{f'(x_n)}$$

Newton's Approximation Method (for Roots of an Equation)

If $f(x) = 0$ has a root between γ and ζ and if $f(\gamma)$ and $f(\zeta)$ have opposite signs, an approximation of the root is given by

$$a = \gamma - \frac{(\gamma - \zeta)\, f(\gamma)}{f(\gamma) - f(\zeta)}.$$

See also **Newton-Raphson Formula.**

Newton's Equation (for Conjugate Distances)

For a perfect lens system, if f and f' are the first and second focal lengths of the system corresponding to space to the left and right of the system respectively, and x and x' are the distances of the object and image from the foci F and F' respectively, as measured outwards from the lens system, then algebraically

$$xx' = ff'$$

Newton's Equation (for the Velocity of Sound)

This equation may be expressed in the form

$$v = \sqrt{\frac{p}{\rho}}$$

where v is the velocity of sound in a gas at pressure p and of density ρ, and is inaccurate as it assumes isothermal conditions.

231

Newton's Identities (for Sums of Powers of Roots)

Relations between the sums of powers of roots of the polynomial equation

$$f(x) = a_0 x^n + a_1 x^{n-1} + \ldots a_r x^{n-r} \ldots a_n = 0$$

$(a_0 \neq 0)$ and its coefficients. If $r_1, r_2, \ldots r_n$ are the roots and

$$s_k = \sum_{r=1}^{n} r^k \text{ and } s_{-k} = \sum_{r=1}^{n} r^{-k}$$

where k is a positive number, then

for $k > n$, $a_0 s_k + a_1 s_{k-1} + \ldots a_r s_{k-r} \ldots a_n s_{k-n} = 0$

for $0 < k \leqslant n$, $a_0 s_k + a_1 s_{k-1} + \ldots a_{k-1} s_1 + k a_k = 0$

for $a_n \neq 0$ and $k > n$, $a_n s_{-k} + a_{n-1} s_{1-k} + \ldots a_{n-r} s_{r-k} +$
$$\ldots + a_0 s_{n-k} = 0$$

and for $a_n \neq 0$ and $0 < k \leqslant n$,

$$a_n s_{-k} + a_{n-1} s_{1-k} + \ldots a_{n-k+1} s_{-1} + k a_{n-k} = 0$$

Newton's Interpolation Formula

If $f(x)$ is given for $x = x_0, x_1, x_2, \ldots x_n$, then $f(x)$ is given approximately by

$$f(x) = f(x_0) + (x - x_0) [x_0 x_1] + (x - x_0)(x - x_1) [x_0 x_1 x_2]$$

$$+ \ldots + (x - x_0) \ldots (x - x_{n-1}) [x_0 x_1 \ldots x_n]$$

where $[x_r x_{r+1}] = \dfrac{f(x_{r+1}) - f(x_r)}{x_{r+1} - x_r}$

and $[x \, x_0 \, x_1 \ldots x_r] = \dfrac{[x_0 \, x_1 \ldots x_r] - [x \, x_0 \, x_1 \ldots x_{r-1}]}{x_r - x}$

Newton's Law of Cooling

The rate at which heat is lost by a heated body is proportional to the difference of temperature between the body and the surrounding medium when the temperature difference is small.

Newton's Law of Gravitation

A particle of mass m_1 attracts a particle of mass m_2 a distance d away with a force

$$F = G \frac{m_1 m_2}{d^2}$$

in the direction of the line joining the particles. If m_1 and m_2 are in gm, d in cm and F in dynes, then $G = 6 \cdot 67 \times 10^{-28}$ c.g.s. units.

Newton's Law of Resistance

At moderate velocities the resistance of a medium is proportional to the square of the velocity.

Newton's Laws of Motion

I. Every body continues in its state of rest or of uniform motion in a straight line except in so far as it may be compelled by impressed force to change that state.

II. Change of motion is proportional to the impressed force and takes place in the direction of the straight line in which the force acts.

III. To every action there is an equal and opposite reaction.

Newton's Method (for Determining the Roots of an Equation)

If x_1 is an approximate value of a root of f$(x) = 0$ then

$$x_1 - \frac{f(x_1)}{f'(x_1)} = x_2 \text{ is a second approximation}$$

$$x_2 - \frac{f(x_2)}{f'(x_2)} = x_3 \text{ is a third approximation}$$

and so on. By repeating the process the root may be found to any degree of approximation and the method can be applied to transcendental as well as algebraic equations providing a point of inflexion does not occur between the first approximation to the root and the true root.

Newton's Rings

When a spherical lens is placed in contact with a plane plate and viewed by reflected light, circular coloured interference fringes are seen

233

surrounding the point of contact, the colours being most brilliant where the air film is thinnest. As the thickness of an air film between a spherical and a plane surface is easily calculated, this phenomenon gave Newton a means of determining accurately the colours produced by air films of any thickness.

Newton's Theory of Lift
The dynamical forces on an aerofoil in a fluid current were explained by Newton as due to the impact of particles of the fluid upon the body This theory, however, is completely unable to explain why an aerofoil should still have positive lift for angles of attack down to $-5°$.

Newton's Theory of Light
Newton enunciated the corpuscular theory of light in 1678. A flight of material particles was emitted by a light source and the impact of these particles on the retina produced the sensation of sight. Rectilinear propagation is a consequence of Newton's second law. Refraction and reflection were explained as due to forces of repulsion and attraction exerted at the surface of the reflecting or refracting medium on the particles.

Newton's 3/8th Rule
A method of computing definite integrals. It is similar to **Simpson's Rule**, except that the interval is divided into $3n$ equal parts giving

$$A = (3/8)\, h\, (y_0 + 3y_1 + 3y_2 + 2y_3 + 3y_4$$

$$+ 3y_5 + 2y_6 + \ldots + 3y_{3n-1} + y_{3n})$$

The notation is defined under **Simpson's Rule**.

Newton-Stirling Interpolation Formula
A modified form of the **Newton-Gauss Interpolation Formula**.

Newton, Trident of
Curve of the form

$$xy = ax^3 + bx^2 + cx + d \quad (a \neq 0)$$

Nicomedes, Conchoid of

A curve whose equation is

$$(x^2 + y^2)(x - a)^2 = x^2 b^2$$

Nietzki's Rule

The colour of a dye is deepened, i.e. the absorption is shifted towards the red, the greater the molecular weight of substituent groups introduced. There are obviously a large number of exceptions to this rule.

Nordheim's Rule

The residual resistivity of a binary alloy containing a mole fraction x of element A and $(1 - x)$ of element B varies as

$$\rho(x) \propto x(1 - x).$$

There are significant deviations from the rule such as when transition metals are alloyed with noble metals.

Nörlund's Definition

$$\log \Gamma(x) = \overset{x}{\underset{0}{S}} \log z \, \Delta z + c$$

where the constant is chosen so that $\log \Gamma(1) = 0$. $\overset{x}{\underset{0}{S}}$ denotes the process of 'summing from 0 to x'. (*See* N. E. Nörlund 'Memoire sur le calcul aux différences finies', *Acta. Math.* 44 (1923), pp. 71–211.)

Norris-Eyring Reverberation Formula

In acoustics, a formula giving the reverberation time T in seconds as

$$T = 0.05 \frac{V}{-S \ln(1 - a_s)}$$

where S = total area of boundaries in sq. ft, V is the volume in cu. ft.

$$a_s = (\overset{n}{\underset{r=1}{\Sigma}} a_r S_r)/S$$

where $a_1, a_2, a_3, \ldots$ = absorption coefficients of areas $S_1, S_2, S_3 \ldots$.

235

Norton's Theorem
A source of electrical energy can be considered as a constant current generator in parallel with an impedance. The current can be considered as the short-circuit current and the impedance is equal to the source impedance.

Nusselt Number
A dimensionless parameter (Nu) used for heat transfer between a solid body and a fluid flowing relative to a body.

$$(Nu) = \frac{h\,l}{\lambda}$$

where h is the heat transfer coefficient (rate of loss of heat per unit area of the surface and per unit temperature difference between the surface and the fluid), l is a characteristic linear dimension of the body and λ the thermal conductivity of the fluid. The **Biot Number** (Bi) is similarly defined but involves the thermal conductivity of the solid. The Nusselt number compares the convective capacity and the conducting capacity of the fluid, whereas the Biot number is an index of the relative resistance to heat flow in the solid and fluid.

Nyquist Rule
If a response-vector locus be plotted for a servo system for positive and negative frequencies and if the point $(-1, j0)$ is not enclosed, the system is unstable.

Nyquist's Theorem
The mean square voltage across a resistance R in thermal equilibrium at absolute temperature T is given by

$$\overline{V^2} = 4\,R\,k\,T\,\Delta f$$

where Δf is the frequency band within which the voltage fluctuations are measured.

O

Obermayer's Test for Indican in Urine
To 4 c.c. of urine and 5 c.c. of Obermayer's reagent (2 gm $FeCl_3$ in
1 litre of pure hydrochloric acid) add 3 c.c. of chloroform. Mix
thoroughly and separate. The chloroform will be blue if indican is
present.

Occam's Razor
A maxim attributed to William (of) Occam (1290–1350): it is vain to
do with more what can be done with fewer, i.e., if the facts resulting
from an experiment can be explained without assuming additional
hypotheses there are no grounds for assuming them.

Ohm's Law
In any circuit consisting of homogeneous conductors through which
electricity is flowing, the ratio of the electromotive force applied across
the circuit to the current flowing in the circuit is a constant if the
circuit is kept under constant physical conditions.

Ohm's Law (Acoustic)
Every motion of air which corresponds to a composite mass of musical
tones is capable of being analyzed into a sum of simple pendular
(harmonic) vibrations, and to each single simple vibration corresponds
a simple tone, sensible to the ear and having a pitch determined by the
periodic time of the corresponding motion of the air.

Olber's Paradox
Assuming that the universe is a static system of infinite age, Olber
deduced that every star must be absorbing as much radiation as it is
emitting and that the bulk of the radiation received by a star or other
body must come from distant parts of the universe. Hence the night
sky should not be dark. The paradox indicates that the universe is
young or is expanding.

Onsager Conductivity Equation (Debye-Hückel-Onsager Equation)
An ion moving in an electrolyte under the influence of an applied

field is retarded by two effects. The relaxation or asymmetry effect is due to the ionic atmosphere of oppositely charged ions having a definite time of relaxation. The ions will also be subjected to the effect of the ionic atmosphere moving in the opposite direction to the central ion. Using these two concepts and making the assumption that the electrolyte is completely dissociated Debye and Hückel were able to deduce an equation for the variation of equivalent conductivity of an electrolyte in any solvent in terms of the concentration. This equation was modified by Onsager;

$$\Lambda = \Lambda_0 - \left\{ \frac{29 \cdot 15(Z_+ + Z_-)}{(\epsilon_r T)^{1/2} \eta} + \frac{9 \cdot 90 \times 10^5}{(\epsilon_r T)^{3/2}} \Lambda_0 \omega \right\} \sqrt{\{C(Z_+ + Z_-)\}}$$

where Z_+ and Z_- are the valencies of the ions, C is the concentration of the electrolyte in gm equivalents/litre, Λ is the equivalent conductivity at this concentration and Λ_0 is the equivalent conductivity at infinite dilution, ϵ_r is the dielectric constant of the solvent and η is its viscosity.

$$\omega = Z_+ Z_- \frac{2q}{1 + \sqrt{q}}$$

where

$$q = \frac{Z_+ Z_-}{Z_+ + Z_-} \frac{\lambda_+ + \lambda_-}{Z_+ \lambda_- + Z_- \lambda_+}$$

and where λ_+ and λ_- are the mobilities of the ions.

Onsager Equation for Dielectric Constant

Onsager has developed an approximate dielectric theory for polar substances. If the induced polarization is neglected the dielectric constant is given by

$$\epsilon = \frac{1}{4}\{1 + 3x + 3(1 + \frac{2}{3}x + x^2)^{1/2}\}$$

$$x = \frac{4\pi n \mu^2}{3kT}$$

where μ is the dipole moment and n is the number of molecules or ions per unit volume.

Onsager's Reciprocal Relations

In the thermodynamics of irreversible processes. This theory states that if a proper choice of the fluxes J_i and forces X_i has been made the matrix of phenomenological coefficients L_{ik} is symmetrical, i.e.,

$$L_{ik} = L_{ki} \quad (i, k = 1, 2, 3, \ldots, n)$$

In irreversible phenomena there can be a number of causes, i.e. a concentration gradient can produce a diffusion flow, a temperature gradient a thermal flow. These causes are called forces or affinities (X) and produce fluxes or flows (J). For situations not too far removed from equilibrium the irreversible phenomena can be expressed by linear relations of the general type

$$J_i = \sum_{k=1}^{n} L_{lk} X_k \quad (l = 1, 2, 3, \ldots, n)$$

Linear relations of this type are called phenomenological relations and the coefficients L_{ik} are called phenomenological coefficients.

Oppenauer Oxidation

When a secondary alcohol is refluxed with aluminium t-butoxide in excess acetone it is oxidized to a ketone

This reaction is specific for secondary alcohols.

Oppenheimer-Philips Reaction

A nuclear 'stripping' process which can occur when a deuteron passes close to a nucleus. The Coulombic force between the nucleus and the deuteron breaks the neutron-proton bond within the deuteron. The neutron is captured by the nucleus and the proton is repelled.

Ostwald Dilution Law

According to W. Ostwald (1888) the dissociation constant of a uni-univalent electrolyte is given by the relation

$$K = \frac{a^2 c}{1 - a}$$

or more accurately by

$$K = \frac{a^2 c}{1 - a} \cdot \frac{f_+ + f_-}{f_\pm}$$

where c is the concentration, a is the degree of dissociation and f_+ and f_- are the activity coefficients of the anions and cations.

Ostwald's Adsorption Isotherm

An expression for the variation of adsorption with gas pressure or concentration

$$a = k.c^{1/n}$$

where a is the quantity adsorbed per unit weight of adsorbent where the concentration of adsorbent in the gas space or the solution is c, and k and n are constants depending on the temperature and nature of adsorbent and adsorbate.

Ostwald's Basicity Rule

W. Ostwald (1887) proposed an empirical relationship between the equivalent conductivity of the sodium or potassium salt of an acid and its basicity

$$\Lambda_{1024} - \Lambda_{32} = 10 \cdot 8 b$$

where Λ_{1024} is the equivalent conductivity of a solution of dilution 1024 and Λ_{32} of dilution 32.

Ostwald Solubility Coefficient

The volume of gas dissolved by a unit volume of solvent at a given temperature and pressure.

Ostwald's Theory of Indicators

All indicators are either weak acids or weak bases whose colour in the ionized form differs from that in the undissociated form.

Otto Cycle
The working cycle of a four-stroke internal combustion engine consisting of inlet, compression, explosion at constant volume and expansion, and exhaust. The whole cycle corresponds to two complete revolutions of the crankshaft.

Overhauser Effect
When placed in an external magnetic field H, a substance which has nuclei of spin ½ and magnetic moment μ_n and also unpaired electrons of spin ½ and magnetic moment μ_e ($\mu_e < 0$) shows this effect. If an r.f. field is applied at the electron spin resonance frequency, then the ratio of the number of nuclei n_+ and n_- with spins 'up' and 'down' in the magnetic field is given by

$$\frac{n_+}{n_-} = e^{2(\mu_n - \mu_e)H/kT} \approx e^{-2\mu_e H/kT}$$

Thus, the polarization of the nuclei is as large as if the nuclei possessed the much larger electron magnetic moment. The effect arises because the nuclei interact predominantly with the electron spins, which in their turn interact thermally with the lattice.

P

Paal-Knorr Synthesis
Pyrrole derivatives may be prepared by treating a $1:4$ diketone with ammonia, a primary amine or a hydrazine.

Paal-Knorr Synthesis of Thiophenes
Homologues of thiophen can be made by reaction of $1:4$ diketones with phosphorus pentasulphide.

$$
\begin{array}{c}
\text{H}_2\text{C} \underline{\hspace{1.5em}} \text{CCH}_2 \\
\mid \qquad \mid \\
\text{RC} \qquad \text{CR} \\
\backslash\backslash \quad /\!/ \\
\text{O} \ \ \text{O}
\end{array}
\ \xrightarrow{\ \text{P}_2\text{S}_5\ }\
\begin{array}{c}
\text{CH} - \text{CH} \\
\| \qquad \| \\
\text{RC} \qquad \text{CR} \\
\backslash \ \ / \\
\text{S}
\end{array}
\ + \text{P}_2\text{O}_2\text{S}_3 + \text{H}_2\text{S}
$$

Paneth's Adsorption Rule
A solid substance strongly adsorbs those radio-isotopes which give rise to insoluble or sparingly soluble components with the acid radical (electronegative component) of the adsorbent. There are many exceptions to this rule, but, as modified by Hahn, the following is strictly true. An element is strongly adsorbed on a precipitate, even from very dilute solution, if the precipitate bears a surface charge of opposite sign to that borne by the element provided the adsorbed compound is very sparingly soluble in the solvent. (*See* **Fajan's Precipitation Rule**.)

Paperitz's Equation
The most general ordinary differential equation of second order with three regular singular points. Let these points be at $z = a,\ b$ and c while the corresponding indices are a, β, γ and a', β', γ' respectively. The equation is

$$
\frac{d^2y}{dz^2} - \left(\frac{a + a' - 1}{z - a} + \frac{\beta + \beta' - 1}{z - b} + \frac{\gamma + \gamma' - 1}{z - c} \right) \frac{dy}{dz}
$$

$$
+ \left\{ \frac{aa'(a - b)(a - c)}{(z - a)^2(z - b)(z - c)} + \frac{\beta\beta'(b - a)(b - c)}{(z - a)(z - b)^2(z - c)} + \frac{\gamma\gamma'(c - a)(c - b)}{(z - a)(z - b)(z - c)^2} \right\}
$$

$$
= 0
$$

where $a + a' + \beta + \beta' + \gamma + \gamma' = 1$. (*See* **Riemann's Symbol**.)

Pappus' Theorems

I. If an arc of a plane curve revolve about an axis in its plane, not intersecting it, the surface generated is equal to the length of the arc multiplied by the length of the path of its mean centre.

II. If a plane area revolve about an axis in its plane, not intersecting it, the volume generated is equal to the area multiplied by the length of the path of its mean centre.

Parkes Process for Desilvering Lead

Base bullion, lead prepared from galena, contains significant quantities of silver. In 1842 Karsten showed that argentiferous lead could be desilvered by the addition of zinc but it was not until 1850 that a practical process was evolved by Parkes. The lead is melted and between 1·5% and 2% of zinc is added. This forms a separate layer on the surface. Silver and gold are more soluble in zinc than in lead, the relative solubility depending of course on the temperature. The zinc layer is separated by a perforated ladle and the process is repeated two or three times with fresh zinc until the percentage of silver in the lead falls to below 5×10^{-3}. The separated material is freed from lead by raising the temperature to the melting point of lead on an inclined plane when the lead melts and flows away and leaves the enriched scum. The zinc is separated by distillation in plumbago retorts and the silver is finally purified by cupellation.

Parseval's Identity

If $\mathbf{i}, \mathbf{j}, \mathbf{k}, \ldots$ are unit orthonormal vectors in a vector space of finite dimensions, then, for any vector $\mathbf{u}$ given by $\mathbf{u} = a\mathbf{i} + b\mathbf{j} + c\mathbf{k} + \ldots$,

$$\| \mathbf{u} \|^2 = | a |^2 + | b |^2 + | c |^2 + \ldots$$

where $\| \mathbf{u} \|$ is the magnitude of $\mathbf{u}$ [given by $+ (\sqrt{\mathbf{u} \cdot \mathbf{u}})$].

Parseval's Theorem

This states that, if, in particular, $F(k)$ and $f(x)$ are Fourier transforms of each other, then

$$\int_{-\infty}^{\infty} |F(k)|^2 \, dk = \int_{-\infty}^{\infty} |f(x)|^2 \, dx$$

Pascal *See* **Appendix**

Pascal's Law
Pressure exerted at any point upon a confined liquid is transmitted undiminished in all directions.

Pascal's Limaçon
When a point (x, y) describes in the z plane a curve L, the parametric representation of which is

$$x = R(\cos \theta + m \cos 2\theta)$$
$$y = R(\sin \theta + m \sin 2\theta)$$

the curve is called Pascal's limaçon.

Pascal's Theorem
For any hexagon inscribed in a conic, the intersections of opposite pairs of sides are collinear (or at infinity).

Pascal's Triangle
A method of obtaining the coefficients of the expansion $(q + p)^n$ (binomial coefficients) from a table;

n	coefficients
1	1 1
2	1 2 1
3	1 3 3 1
4	1 4 6 4 1

Each term is derived from the sum of the two terms lying on either side of it in the line above.

Paschen-Back Effect
Paschen and Back discovered in 1912 that all the simplicity of many Zeeman patterns was lost when the magnetic separation was comparable to the natural separation of a doublet, triplet, etc. The lines appeared to influence each other until, when the magnetic separation was large compared to the natural separation, the normal **Zeeman Effect** returned both in pattern and magnitude.

A similar effect in which the hyperfine structure in a spectrum breaks up in a magnetic field but begins to overlap again as the field is further increased is called the **Back-Goudsmit Effect**.

Paschen Series *See* **Balmer Series**

Paschen's Law
At a constant temperature, the breakdown voltage of a gas is a function only of the product of the gas pressure and the distance between parallel plane electrodes.

Pattinson Process for Desilvering Lead
In 1833 H. L. Pattinson observed that on heating a bar of lead containing a small proportion of silver the first drops which oozed out were richer in silver than the bar and conversely, for lead bars with a high proportion of silver, the first drops of metal contained less silver than the residual lead. This situation is easily understood with reference to the phase diagram,

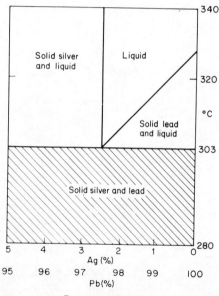

Pattinson Process

Pauli Exclusion Principle
If individual systems belong to a certain class, e.g. electrons, neutrons, protons, only antisymmetric functions may be used to describe the assemblage. This principle, in particular, states that it is impossible for any two electrons in the same atom to have four identical quantum numbers.

Pauli Spin Matrices
These are denoted

$$\sigma_x = \begin{pmatrix} 0 & 1 \\ 1 & 0 \end{pmatrix} \qquad \sigma_y = \begin{pmatrix} 0 & -j \\ j & 0 \end{pmatrix} \qquad \sigma_z = \begin{pmatrix} 1 & 0 \\ 0 & -1 \end{pmatrix}$$

$$\sigma^2 = \begin{pmatrix} 3 & 0 \\ 0 & 3 \end{pmatrix} \text{ where } \sigma_x^2 + \sigma_y^2 + \sigma_z^2 = \sigma^2.$$

The total spin (intrinsic angular momentum) of an electron is then expressed as $S = \frac{1}{2}\hbar\sigma$.

Pauli Spin Susceptibility
The paramagnetic susceptibility of a free electron gas. It arises from the ability of electrons within an energy of approximately kT of the **Fermi Energy** to alter their direction of spin in a magnetic field.

Peano's Axioms
These axioms define the properties of positive integers;
(1) 1 is a positive integer.
(2) Each positive integer n has a unique successor denoted $S(n)$.
(3) $S(n) \neq 1$.
(4) If $S(n) = S(m)$ then $n = m$.
(5) The principle of finite induction holds; i.e. the set of positive integers 1 plus the successors $S(n)$ for each n contains all positive integers.

Pearl-Reed Curve
Curve of the form

$$y = k/(1 + e^{a+bx}) \text{ where } b < 0.$$

Pearson's Distributions
Frequency functions $\phi(x)$ of many continuous distribution functions, for instance **Gauss' Error Curve**, can be expressed in the form

$$\frac{1}{\phi}\frac{d\phi}{dx} = \frac{x - a}{b_0 + b_1 x + b_2 x^2}.$$

The parameters a, b_0, b_1 and b_2 define each distribution and the distributions are classified according to the roots of the equation

$$b_0 + b_1 x + b_2 x^2 = 0.$$

Peclet's Number
A dimensionless number for a fluid flowing past a body and equal to
Prandtl's Number × Reynolds Number.

Pellian Equation
Equation of the form

$$x^2 - Ay^2 = 1$$

where A is a positive integer that is not a perfect square.

Peltier Effect
Heat is, in general, absorbed or liberated when an electric current
crosses the junction between two metals A and B. The heat is
proportional to the current i and is equal to Πi, where Π is called the
Peltier coefficient of A with respect to B and varies with temperature.

Penning Effect
The ionization of one species of gas molecule by collision with another
species in a mixture of gases, when the former species is in a metastable
excitation level which is above the ionization potential of the other.

Perkin Reaction
When an aromatic aldehyde is heated with an anhydride of an aliphatic
salt containing two a-hydrogen atoms in the presence of its sodium
salt condensation takes place to form a β-aryl acrylic acid.

CHO + CH₃COONa   CH=CH.COOH

⬡   Acetic anhydride   ⬡   + H₂O
                  ⟶

Perkin's Phenomenon
The decrease which occurs in the conductivity of carbon when
negatively electrified.

Pettenkofer's Test
A test for bile depending upon the bile salts. A red coloration is
obtained when the bile is mixed with sucrose and concentrated
sulphuric acid.

Petzval Surface

The paraboloidal surface over which an image is formed by a lens system in the absence of astigmatism.

Pfaffian Differential Equation

A first-order differential equation of form

$$A(x, y, z)dx + B(x, y, z)dy + C(x, y, z)dz = 0.$$

It is integrable if, and only if, there is an integrating factor $\mu = \mu(x, y, z)$ such that $\mu(A dx + B dy + C dz)$ is an exact differential.

Pfitzinger Reaction

Derivatives of cinchonic acid (4 quinoline carboxylic acid) may be prepared by the reaction of isatin with aldehydes and ketones.

Pfund Series *See* **Balmer Series**

Picard's Method

A method of solving ordinary differential equations by successive approximations. If

$$dy/dx = f(x, y)$$

a solution may be found of the form

$$y = y_0 + \int_{x_0}^{x} f(x, y)dx \equiv y_0 + \int_{x_0}^{x} \frac{dy}{dx}\ dx$$

where $y = y_0$ at $x = x_0$. If the approximation $y = y_0$ under the integral sign is made, the integral is suitable for numerical integration as it is a function of x and y_0 only. Thus,

$$^1y = y_0 + \int_{x_0}^x f(x, y_0)\, dx$$

The process is repeated to give

$$^2y = y_0 + \int_{x_0}^x f(x, {}^1y)\, dx$$

and so on.

Pinkhof Titration
An electrochemical titration method wherein the reference electrode is set so that it assumes the potential shown by the indicator at the end-point.

Pippard Coherence Length
The range of coherence, ζ, of the wave-functions of superconducting electrons in a superconductor. Its value at zero temperature is taken as

$$\zeta_0 = \frac{h\, v_F}{\pi^2 E_g}$$

where E_g is the energy gap obtained from the theory of **Bardeen, Cooper and Schrieffer** and v_F is the velocity of electrons possessing the **Fermi Energy**. In tin, ζ is approximately 10^{-4} cm. In applying the **London Equation,** if the vector potential is varying rapidly with position, it is necessary to consider the current density at a point as proportional to an average of the vector potential over a region of characteristic size ζ.

Planck's Constant *See* Appendix

Planck's Quantum Theory

The Rayleigh-Jeans formula for the distribution of energy in the spectrum of black body radiation is:

$$E = \frac{8\pi kT}{\lambda^4}$$

indicating a steady increase of energy density E with decreasing wavelength λ as for the dotted curve in the figure. The experimental curve is shown by the full line and has a maximum for a given value of λ depending upon the temperature of the black body. This discrepancy

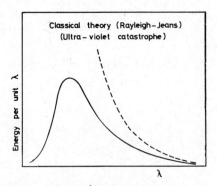

Planck's Quantum Theory

was resolved by Planck who showed that the error lay in applying the classical laws of mechanics to atomic particles. These observations were the basis of the modern quantum theory. In classical theory the amplitude of vibration of a linear oscillator may have any desired value within certain limits. Planck assumed, however, that the amplitude of vibration could only assume certain discrete values.

Planck's Radiation Formula *See* Planck's Quantum Theory

Playfair's Axiom

Two intersecting straight lines cannot be parallel to a third straight line.

Pockel's Effect

If a light beam is passed through certain crystals in the same direction as an applied electric field, the refractive indices of the crystals change with change of applied field and the crystals become birefringent with the difference in velocity for the two rays being proportional to the electric field. If a plane polarized beam of light is used, the plane of polarization is rotated as the light passes through the crystal. The effect is analogous to the **Faraday Effect** in a magnetic field. It has been widely used in solid-state optical shutters and light modulators; important electro-optic devices. Practically, the most important materials for constructing such devices are ADP (ammonium dihydrogen phosphate), EDT (ethylamine dihydrogen tartrate) and zinc sulphide. *See also* **Kerr Electro-optic Effect**.

Poise *See* **Appendix**

Poiseuille's Formula
The volume of liquid V, flowing per second through a capillary tube of length l and radius r under a pressure p is given by

$$V = \frac{\pi p r^4}{8 l \eta}$$

where η is the viscosity of the liquid.

Poisson Distribution Law
The frequency distribution for the number of random events r in a constant time interval is given by

$$P(r) = \frac{m^r e^{-m}}{r!}$$

where m is the mean number of counts in the time interval.

Poisson's Bracket
A differential operator on two (twice continuously differentiable) functions U and V describing a physical state:

$$[U, V] = \sum_{m=1}^{n} \left[\frac{\partial U}{\partial p_m} \frac{\partial V}{\partial q_m} - \frac{\partial U}{\partial q_m} \frac{\partial V}{\partial p_m} \right]$$

where p and q are conjugate variables, e.g. momenta and coordinates (cf. **Hamiltonian**) and n is the number of degrees of freedom of the system.

The **Heisenberg Uncertainty Principle** can be stated thus: If the Poisson bracket $[U, V] = K$ where K is a constant independent of the coordinate system used, the product of the uncertainties in simultaneous measurement of U and V, i.e. ΔU and ΔV respectively, is given by

$$\Delta U \, \Delta V = Kh$$

where h is Planck's constant.

If W is a similar function depending on p and q, then

$$\left[U, [V, W] \right] + \left[V, [W, U] \right] + \left[W, [U, V] \right] = 0$$

and this is called **Poisson's Identity**.

Poisson's Equation
If, within a closed surface, there exists a volume density of charge ρ the potential V on the surface will be given by

$$\nabla^2 V = \frac{\partial^2 V}{\partial x^2} + \frac{\partial^2 V}{\partial y^2} + \frac{\partial^2 V}{\partial z^2} = 4\pi\rho$$

This determines the distribution of charge within a surface when the potential at every point on the surface is known.

Poisson's Integral Formula (for Bessel Functions)
If $J_n(x)$ is a **Bessel Function**, then

$$J_n = \frac{2 \, (x/2)^n}{\sqrt{\pi} \, \Gamma \, (n + \frac{1}{2})} \int_0^{\pi/2} \cos \, (x \cos t) \sin^{2n}t \; dt$$

$$(n > -\tfrac{1}{2})$$

Poisson's Law
When a gas expands adiabatically
$$PV^\gamma = \text{constant}$$
where P is the pressure, V is the volume and $\gamma = C_p/C_V$ the ratio of the specific heats of a gas.

Poisson's Ratio
The ratio of lateral strain to longitudinal strain when a body is in simple tension or compression.

Poisson's Sum Formula
A method of evaluating a sum of the form

$$\sum_{n=-\infty}^{\infty} f(an)$$

Let $F(k)$ be the **Fourier Transform** of $f(x)$. The sum is equal to

$$\frac{\sqrt{(2\pi)}}{a} \sum_{n=-\infty}^{\infty} F\left(\frac{2n\pi}{a}\right)$$

which may be easier to evaluate.

Polonovski Reaction
The removal of a methyl group can be accomplished by this reaction.
A tertiary amine is converted to the amine oxide. This compound reacts
with acetic anhydride to give formaldehyde and an amide which can be
hydrolysed to the secondary amine.

Portevin-Le Chatelier Effect
The continually repeating non-smooth deformation of a specimen when
subjected to a uniformly increasing stress. In polycrystalline material
the initial steps are due to yielding inside the single crystal grains but
the later steps are due to the presence of grain boundaries.

Pourbaix Diagram
In any electrode reaction, electrode potentials can be plotted against
pH. If this is done for various activities of the reactants a diagram can
be constructed in which different areas or domains can be associated
with different behaviours of the reactants. Thus for the behaviour of
iron in aqueous solution a diagram
can be constructed as shown where
the domains correspond to

A. Passivation if the oxide stable in
this area is protective.
B. Corrosion. Iron ions are stable in
this area.
C. Immunity due to iron becoming
cathodic.
D. Corrosion due to formation of
iron anions.

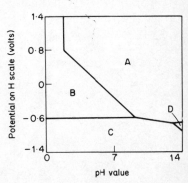

Poynting's Theorem
When energy is transmitted through an electromagnetic field, if a small volume is enclosed by a surface in the field, and if the energy of electric strain and magnetic flux contained in it be varying, the amount of energy which enters each element of the surface is measured by the sum of the products of the electric and magnetic forces resolved along each element of the surface, multiplied by the sine of the angle between their directions and divided by 4π.

The vector measuring the rate of flow of electromagnetic energy per unit area is known as **Poynting's Vector**.

Prandtl's Number
A dimensionless number (Pr) for a fluid defined as

$$(Pr) = \frac{C_p \, \eta}{\lambda \rho}$$

where η is the dynamic viscosity of the fluid, C_p is its specific heat at constant pressure, λ its thermal conductivity and ρ the density.

Preston's Rule
In the anomalous **Zeeman Effect**, lines of the same series show the same pattern.

Prévost Theory of Heat Exchange
In the eighteenth century ideas regarding radiant energy were confused. Radiations were described as hot and cold, e.g. a block of ice emitted cold radiation. Prévost recognized the looseness of this method of description in 1792, and said that all bodies emit radiant energy, the amount of which increases with temperature and is not affected by surrounding bodies. The rise and fall of temperature of a body is due to the exchange of radiant energy with other surrounding bodies.

Prileschaiev Reaction
Peracids convert olefines into olefine oxides.

$$RCH{=}CHR + C_6H_5COO_2Na \rightarrow R.CH\underset{\displaystyle \diagdown \;\; \diagup}{\overset{\displaystyle O}{}}CH.R + C_6H_5COONa$$

Usually perbenzoic or monoperphthalic acids are used in this reaction.

Pringsheim's Theorem (for a Convergent Series)
If Σu_n converges and $u_n \to 0$ monotonically, then $nu_n \to 0$.

Prins Reaction
Arylethylenes react with paraformaldehyde in acetic acid solution in the presence of concentrated sulphuric acid to give unsaturated alcohols, glycols or their acetates.

Proca Equation
A relativistic equation for a form of vector potential describing a hypothetical particle of unit spin. See **Dirac's Equation** and **Klein-Gordon Equation**.

Procopiu Effect
The appearance in a coil wound coaxially around a ferromagnetic wire of an electromotive force which has double the frequency of an alternating current passed through the wire.

Prout's Hypothesis
W. Prout in 1815 suggested that all atoms were built up from the same unit, the hydrogen atom. The fact that not all atomic weights were integral led to this hypothesis being regarded as unsound. The discovery of isotopes removed this objection and a modified form of the hypothesis utilizing the proton or hydrogen nucleus and the neutron as the building units of all atomic nuclei has gained universal acceptance.

Prym's Functions

$$P(x;\rho) = \int_0^\rho t^{x-1} e^{-t} \, dt$$

and

$$Q(x;\rho) = \int_\rho^\infty t^{x-1} e^{-t} \, dt$$

are known as Incomplete Gamma Functions. The special functions which arise when $\rho = 1$ are called Prym's Functions.

Pschorr Synthesis

A method of synthesizing phenanthrene. *o*-nitro-benzaldehyde is
condensed with sodium phenylacetate by means of the **Perkin Reaction.**
The product *a*-phenyl *o*-nitrocinnamic acid is reduced and diazotised.
The azo-compound is treated with sulphuric acid and copper powder
to give phenanthrene-9-carboxylic acid. This compound loses carbon
dioxide when heated strongly and phenanthrene is formed,

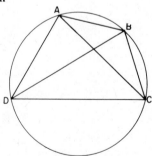

This synthesis offers means of preparing substituted phenanthrenes.

Ptolemy's Theorem

$$AC \cdot BD = AD \cdot BC + AB \cdot CD$$

Ptolemy's Theorem

The rectangle contained by the diagonals of a quadrilateral inscribed in a circle is equal to the sum of the two rectangles contained by its opposite sides.

Purdie Method of Alkylation
When an alkyl halide is heated with an alcohol in the presence of silver oxide an ether is formed with the elimination of water and silver iodide.

Purkinje Effect
The human eye is more sensitive to blue light than to yellow light when illumination is poor (less than 0·1 lumens per sq. ft) and to yellow rather than blue when the illumination is good.

Pythagoras' Theorem
In a right-angled triangle the square on the hypotenuse is equal to the sum of the squares on the other two sides.

Pythagorean Identity
The trigonometric identity

$$\sin^2 x + \cos^2 x = 1.$$

Pythagorean Numbers
Any set of integers satisfying the equation $x^2 + y^2 = z^2$. They are given by $a^2 - b^2$, $2ab$ and $a^2 + b^2$, where a and b are any integers.

Q

Quincke Effect

A substance of high magnetic susceptibility tends to move into the strong region of a magnetic field. Similarly, a substance of high dielectric constant tends to move into a strong electric field.

R

Raabe's Test for Convergence

The positive term series Σu_n is convergent if

$$n\left(\frac{u_n}{u_{n+1}} - 1\right) \geqslant \rho \text{ where } \rho > 1.$$

It is divergent if

$$n\left(\frac{u_n}{u_{n+1}} - 1\right) \leqslant 1.$$

Raman Scattering

If any substance, gaseous, liquid or solid is exposed to radiation of a definite frequency, the light scattered at right angles to the incident beam contains a spectral distribution of frequencies characteristic of the substance under investigation. Such a spectrum is known as the **Raman Spectrum**. It arises because of interaction with molecular vibrations in the substance. A most important phenomenon recently observed is coherent Raman scattering. It has been observed that if certain substances are excited by coherent radiation from a laser, the Raman spectrum contains coherent components.

Ramsauer Effect

For low energy electrons passing through a gas, there is a peak in the absorption cross-section at a particular electron energy. The peak corresponds to an energy at which electrons are particularly efficient at ionizing or exciting atoms in the gas. Ramsauer first discovered the effect in the gases xenon, krypton and argon.

Ramsay-Young Rule

If two substances have the same vapour pressure p at temperature T_A

and T_B and p' at T_A' and T_B' respectively, Ramsay and Young in 1885 showed empirically that,

$$\frac{T_A}{T_B} - \frac{T_A'}{T_B'} = C(T_A - T_A')$$

where C is a constant.

Rankine Cycle

A steam engine cycle, characterized by the introduction of water at boiler pressure by a pump, evaporation, adiabatic expansion to condenser pressure and condensation to the initial point.

Rankine Efficiency

The efficiency of an ideal engine working under a **Rankine Cycle**.

Rankine-Hugoniot Relations

Equations expressing the conservation of mass, momentum and energy on either side of the discontinuity in a shock wave.

$$\rho_1 v_1 = \rho_2 v_2$$
$$p_1 + \rho_1 v_1^2 = p_2 + \rho_2 v_2^2$$
$$\tfrac{1}{2}v_1^2 + U_1 + \frac{p_1}{\rho_1} = \tfrac{1}{2}v_2^2 + U_2 + \frac{p_2}{\rho_2}$$

Subscripts 1 and 2 refer to fluid immediately to the right and left of the discontinuity respectively. p is pressure, ρ density, v flow velocity and U internal energy. The **Hugoniot function** $\mathcal{H}$ is defined by

$$\mathcal{H} = U_2 - U_1 - \tfrac{1}{2}\left(p_2 + p_1\right)\left(\frac{1}{\rho_1} - \frac{1}{\rho_2}\right)$$

and by elimination of the flow velocities from the Rankine-Hugoniot relations it can be shown that where these equations are applicable the Hugoniot function is zero.

Rankine's Formula

An empirical formula for the collapsing load for a column

$$L = \frac{\sigma A}{1 + a(l/k)^2}$$

where L = the collapsing load,

A = area of cross-section,

σ = safe compressive stress,

l = length of column,

$a = \sigma/\pi^2 E$ where E is Young's Modulus,

k = least radius of gyration of section.

Rankine Temperature Scale

A scale of temperature with the freezing point of water as $491 \cdot 7°$ and the boiling point of water as $671 \cdot 7°$ at normal pressure.

Raoult's Law

If p_0 is the vapour pressure of a pure solvent and p the vapour pressure of a substance containing n_2 moles of solute in n_1 moles of solvent. F. Raoult showed that

$$\frac{p_0 - p}{p} = \frac{n_2}{n_1 + n_2} = x$$

where x is the mole fraction of solute.

Raschig Process

Phenyl chloride is produced commercially by passing benzene vapour, air and hydrogen chloride over cuprous chloride as catalyst.

$$C_6H_6 + HCl + \tfrac{1}{2}O_2 \xrightarrow{Cu_2Cl_2} C_6H_5Cl + H_2O$$

Rayleigh Criterion

Two patterns (for instance **Fraunhofer Diffraction** patterns or **Airy's Discs**) of equal intensity may be said to be resolved when the central maximum of one pattern falls over the minimum of the other.

Rayleigh-Jeans Radiation Formula *See* **Planck Quantum Theory**

Rayleigh-Ritz Method

A variational method of obtaining approximate solutions of differential equations by assuming arbitrary parameters in trial functions and continuously improving them by iteration.

Rayleigh Scattering

A process of scattering of electromagnetic radiation in which there is no change of wavelength and in which the intensity of the scattered radiation (usually in the optical region) is proportional to ν^4. If the scattering centres can be represented by dipole oscillators of natural frequency ν_0, then, if $\nu \ll \nu_0$, the intensity of the scattered light is given by

$$I_s = \frac{8 \pi N e^4 \nu^4}{3m^2 c^4 \nu_0^4} I_0$$

where I_0 is the intensity of the primary radiation and N is the number of oscillators of mass m and charge e. If $\nu \gg \nu_0$, **Thomson Scattering** occurs.

Alternatively, if the scattering is considered to be by N dielectric spheres of radius r where $r \ll \lambda$ (λ the wavelength of the light), then the intensity of the scattered light at an angle θ to the incident radiation and distance d from the scattering centres is given by

$$I_s = \frac{9}{2} \frac{\pi^2 N}{d^2} \left(\frac{n^2 - 1}{n^2 + 2}\right)^2 \frac{V^2}{\lambda^4} (1 + \cos^2 \theta) I_0$$

where n is the refractive index of the spheres relative to the surrounding medium and V is the volume of the spheres. *See also* **Tyndall Effect**.

Rayleigh's Equation of Group Waves

$$\nu' = \nu - \lambda \frac{d\nu}{d\lambda}$$

This relation is known as Rayleigh's Equation where ν' is the group velocity of a wave group, ν is the wave velocity and λ is the wavelength.

Rayleigh Waves

In a solid elastic medium of finite size, a disturbance will produce surface waves in addition to waves moving through the bulk material. Rayleigh waves are surface waves whose vibrations are in a plane vertical to the surface and containing the direction of propagation. Where the vibrations are horizontal and perpendicular to the direction of propagation, the waves are called **Love Waves**. Surface waves are attenuated less with distance than waves propagated through the bulk material and are of particular importance, therefore, at large distances from the source.

Réaumur Scale

A scale of temperature with $0°$ at the freezing point and $80°$ at the boiling point of water at normal pressure. $0°R = 0°C$, $80°R = 100°C$.

Reformatsky Reaction

An a- or β-bromacid ester will react with a carbonyl compound in the presence of zinc to form a β- or γ-hydroxy ester. The reaction is carried out by adding zinc to a mixture of the bromacid ester and the carbonyl compound.

$$CH_2Br.COOC_2H_5 \xrightarrow{Zn} Br.Zn.CH_2.COOC_2H_5$$

$$\downarrow CH_3CHO$$

$$\underset{\underset{OH}{|}}{CH_3CH.CH_2COOC_2H_5} \xleftarrow[\text{acid}]{\text{dilute}} \underset{\underset{OZnBr}{|}}{CH_3CH.CH_2COOC_2H_5}$$

Reichert-Meissl Value

The number of millilitres of $0·1N$ potassium hydroxide required to neutralize 5 gm of hydrolysed fat is known by this name and indicates the amount of steam volatile fatty acids (up to lauric acid) present in the fat.

Reimer-Tiemann Reaction

o-hydroxybenzaldehyde contaminated with a little of the p-isomer, may be prepared by refluxing an alkaline solution of phenol with chloroform. The excess chloroform is distilled off and the residual liquid acidified.

263

Replacement of the chloroform by carbon tetrachloride leads to a phenolic acid instead of the aldehyde. The mechanism outlined above is not necessarily the correct one but indicates one possible course of the reaction.

Reynolds Number
When a number of geometrically similar bodies fall through a fluid of viscosity η and density ρ with a velocity v, the Reynolds number is, for any one of the bodies, $R = \rho l v / \eta$ where l is the length of a corresponding linear dimension. The specific resistance to motion for a number of geometrically similar bodies in a fluid is the same in all cases at velocities for which the Reynolds number has the same numerical value.

Riccati-Bessel Functions
The differential equation

$$x^2 \, (d^2z/dx^2) + \left[x^2 - n(n + 1) \right] z = 0$$

$(n = 0, \pm1, \pm2, \ldots)$ has a general solution of the form;
$$z = S_n(x) + C_n(x)$$
$$= \sqrt{(\pi x/2)} \, J_{n+\frac{1}{2}}(x) + \sqrt{(\pi x/2)} \, N_{n+\frac{1}{2}}(x)$$

where $S_n(x)$ and $C_n(x)$ are independent solutions called Riccati-Bessel functions. See also **Bessel Functions**.

Riccati's Equation
A first-order nonlinear differential equation of the form;

$$dy/dx = f_1(x)y^2 + f_2(x)y + f_3(x)$$

Richardson-Dushman Equation
In the emission from a metallic thermionic cathode

$$i = A(1 - r)T^2 e^{-b/T}$$

where i = saturation current density (amp cm^{-2})
A = a constant equal to 120 amp cm^{-2} $^{\circ}$K^{-2}
b = absolute temperature equivalent of the work function,
r = a reflection coefficient to allow for irregularities of the surface.

Richardson Effect *See* **Barnett Effect**

Richards' Rule
The molar heat of fusion of a solid divided by its melting point is equal
to approximately 2. The rule is analogous to **Trouton's Law** for
vaporization but is less generally applicable.

Rieke Diagram
A graph in polar coordinates showing the behaviour of electronic tubes
as a function of the load impedance.

Riemann-Christoffel Tensor *See* **Christoffel Symbols**

Riemann's Lemma
If $\Phi(x)$ is non-decreasing and bounded in the range $b > x > a$, and λ is
large, then

$$\int_a^b \Phi(x) \cos(\lambda x)dx \text{ and } \int_a^b \Phi(x) \sin(\lambda x)dx$$

are $O(1/\lambda)$.

Riemann's Surfaces
A geometrical representation of multi-valued functions in the complex
plane which is developed into two or more interleaving surfaces.

Riemann's Symbol
A shorthand notation for a general solution of an ordinary
differential equation with three regular singular points. For the
Paperitz's Equation this takes the form

$$y = P \begin{pmatrix} a & b & c \\ \alpha & \beta & \gamma & z \\ \alpha' & \beta' & \gamma' \end{pmatrix}$$

The top three terms (a, b, c) give the positions of the regular singular
points in the complex z plane. The other two rows give corresponding
values of the indices (i.e. solutions of the indicial equation). Thus at a,
the indices are α and α'. The z in the fourth column denotes the
independent variable.

Riemann Zeta Function

Defined by either

$$\zeta(z) = \sum_{n=1}^{\infty} n^{-z}$$

or

$$\zeta(z) = \frac{1}{\Gamma(z)} \int_0^{\infty} \frac{x^{z-1}}{e^x - 1} \, dx$$

where $z = x + jy$ and $x > 1$.

Righi-Leduc Effect

If a temperature gradient dT/dx is established along a conductor or semiconductor and if a magnetic field is applied orthogonal to this temperature gradient, then a further temperature gradient dT/dy is established at right angles to both the magnetic field and to the first temperature gradient, such that

$$\frac{dT}{dy} = S \, B_z \, \frac{dT}{dx}$$

where S is the Righi-Leduc coefficient and B_z is the magnetic flux density.

Rinmann's Green Reaction

A reaction for zinc based on the fact that, when cobalt oxide is ignited with zinc oxide, a green spinel type double oxide is formed.

Ritter Reaction

Tertiary carbonium ions formed from alkenes or alcohols add to nitriles to give nitrilium salts. Addition of water to this salt gives a tertiary alkyl amide.

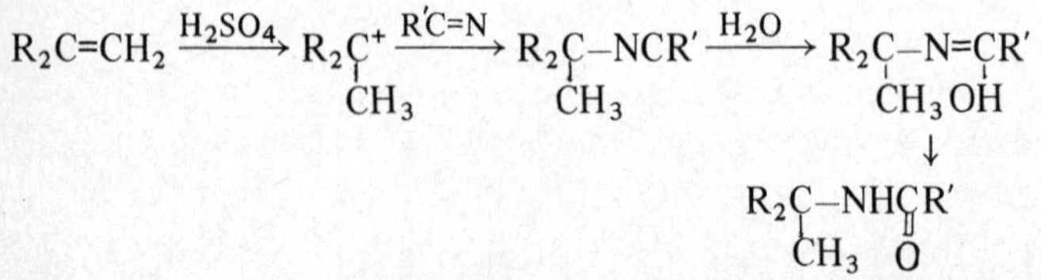

$$R_2C{=}CH_2 \xrightarrow{H_2SO_4} \underset{CH_3}{R_2C^+} \xrightarrow{R'C{\equiv}N} \underset{CH_3}{R_2C{-}NCR'} \xrightarrow{H_2O} \underset{CH_3 OH}{R_2C{-}N{=}CR'}$$

$$\downarrow$$

$$\underset{CH_3 \quad O}{R_2C{-}NHCR'}$$

Ritz Combination Principle

The wave number of any spectral line in a given series may be represented by the difference of two terms.

$$\bar{\nu} = \frac{R}{x^2} - \frac{R}{y^2}$$

x remains constant for any given series and y assumes different integral values to give the lines in that series. R is a constant known as the **Rydberg Constant.**

Robert's Law

For every mechanical linkage there are at least two substitutes that will produce the same desired motion. Furthermore, the two alternative linkages are related to the first by a series of similar triangles.

Rodriques Formula

$$P_m(x) = \frac{1}{2^m m!} \frac{d^m}{dx^m} (x^2 - 1)^m$$

P_m is the **Legendre Coefficient.**

Rolle's Theorem

Suppose $f(x)$ is continuous in the closed interval (a, b) and has a derivative $f'(x)$ for every x such that $a < x < b$.

If $f(a) = 0$ and $f(b) = 0$, then $f'(x) = 0$ for at least one value of x between a and b.

Röntgen *See* Appendix

Rosenmund Reduction

Aldehydes may be prepared by the reduction of an acid chloride with hydrogen in boiling xylene using a suspension of palladized barium sulphate as catalyst. Reduction of the aldehyde to the alcohol does not occur as barium chloride inhibits this reaction.

Rothera's Test (for Acetone and Acetoacetic Acid in Urine)
Saturate 10 c.c. of urine with ammonium sulphate and then add 2–3 drops of freshly prepared 5% sodium nitro-prusside. Mix and allow to

stand. A deep permanganate colour will develop in the presence of acetone and acetoacetic acid. The test is not negative until the solution has stood for ten minutes.

Roth's Theorem

If ζ is any algebraic number that is a root of an equation

$$a_0 u^n + a_1 u^{n-1} + \ldots + a_n = 0$$

where the a's are integers, then there are only a finite number of pairs of integers p, q such that

$$\left| \zeta - \frac{p}{q} \right| < \frac{1}{q^k}$$

if $k > 2$. (*Nature*, 13.12.58, p. 1629.)

Routh's Rule

A rule for finding the roots of an equation, whose real parts are positive, by inspection of variations of signs of matrices of its coefficients. If $f(x) = a_0 x^n + a_1 x^{n-1} + \ldots a_{n-1} x + a_n = 0$ is the equation and

$$M_0 = a_0 > 0 \qquad M_1 = a_1 \qquad M_2 = \begin{vmatrix} a_1 & a_0 \\ a_3 & a_2 \end{vmatrix}$$

$$M_3 = \begin{vmatrix} a_1 & a_0 & 0 \\ a_3 & a_2 & a_1 \\ a_5 & a_4 & a_3 \end{vmatrix} \qquad M_4 = \begin{vmatrix} a_1 & a_0 & 0 & 0 \\ a_3 & a_2 & a_1 & a_0 \\ a_5 & a_4 & a_3 & a_2 \\ a_7 & a_6 & a_5 & a_4 \end{vmatrix}$$

and so on, then the number of roots equals the number of sign changes in one of the sequences

$$M_0, M_1, M_2/M_1, M_3/M_2, \ldots M_n/M_{n-1}$$

or

$$M_0, M_1, M_1 M_2, M_2 M_3, \ldots M_{n-1} M_{n-2}, a_n.$$

268

Routh's Rule of Inertia

The moment of inertia I of a body about an axis of symmetry is given by;

$$I = \frac{M(a^2 + b^2 + c^2)}{d}$$

where M is the mass and a, b and c the lengths of the perpendicular semi-axes of the body. $d = 3, 4$ or 5 according to whether the body is a rectangular parallelopiped, an elliptic cylinder or an ellipsoid.

Rowland Circle

A circle having the radius of curvature of a concave diffraction grating as diameter. If a slit is placed anywhere on the circumference of the circle, spectra are formed in exact focus also on the circumference.

Ruff Degradation

An aldohexose may be converted to the corresponding aldopentose by first oxidizing it to the aldonic acid with bromine water, converting the acid to the calcium salt and treating it with Fenton's Reagent.

$$
\begin{array}{ccc}
\text{CHO} & \text{COOH} & \\
| & | & \\
\text{CHOH} & \text{CHOH} & \text{CHO} \\
| \quad \xrightarrow[\text{H}_2\text{O}]{\text{Br}_2} & | \quad \xrightarrow[\text{H}_2\text{O}_2 + \text{Fe}^{++}]{\text{Ca salt}} & | \\
(\text{CHOH})_3 & (\text{CHOH})_3 & (\text{CHOH})_3 + \text{CO}_2 \\
| & | & | \\
\text{CH}_2\text{OH} & \text{CH}_2\text{OH} & \text{CH}_2\text{OH}
\end{array}
$$

Ruggli Principle

According to this principle by using sufficiently dilute solutions of hydroxyacids the distance between the different molecules can be made greater than the distance between the hydroxyl group and carboxyl group in the same molecule. The formation of a cyclic lactone will be preferred, therefore, to the condensation of two molecules together to form long chains.

Runge-Kutta Method (of Numerical Integration)

A widely used method of computing definite integrals in which fixed increments h of the independent variable x are considered. For a first-order equation of the form

$$y' = f(x, y)$$

the following are obtained;

$k_1 = f(x_0, y_0)h$ where x_0 and y_0 are initial values of x and y
$k_2 = f(x_0 + h/2, y_0 + k_1/2)h$
$k_3 = f(x_0 + h/2, y_0 + k_2/2)h$
$k_4 = f(x_0 + h, y_0 + k_3)h.$

This gives values of x and y, after the first increment of

$$x_1 = x_0 + h \text{ and } y_1 = y_0 + \Delta y$$

where $\Delta y = 1/6(k_1 + 2k_2 + 2k_3 + k_4).$

Values of x and y after further increments are obtained in a similar way. The error is $O(h^5)$. The method can also be applied to second-order equations and to simultaneous differential equations.

Runge's Law

The difference of the wave-numbers of the limits of the diffuse and **Bergmann Series** in alkali spectra is equal to the wave-number of the first line of the diffuse series.

Runge's Rule

In the complex Zeeman effect, in a given magnetic field the frequency shift of a line is either that of a Balmer line $(e/4\pi mc)H$ (*see* **Zeeman Effect**) or m/n times this, where m and n are small integers.

Russell-Saunders Coupling

A method of explaining atomic spectra in which the orbital momenta of the electrons are assumed to be strongly coupled because of strong electrostatic forces and the electron spins are strongly coupled as a consequence of the **Pauli Exclusion Principle**. For the lighter elements, Russell Saunders coupling is a good approximation.

Rutgers Equation
In the theory of superconduction

$$\left(\frac{dH_T}{dT}\right)_{T=T_0} = \sqrt{\left[\frac{4\pi}{VT_0}(C^s - C^n)_{T=T_0}\right]}$$

H_T is the magnetic field intensity at which the superconductor loses its superconductivity. T_0 is the temperature at which metal becomes superconducting. C^s and C^n are the specific heats in the superconducting and normal state.

Rutherford *See* **Appendix**

Rutherford Scattering
Scattering by heavy nuclei of light charged particles. The probability of scattering into a solid angle lying between ω and $\omega + d\omega$ is given by the differential scattering cross-section $d\sigma$;

$$d\sigma = \frac{b^2}{16 \sin^4 \theta/2} \, d\omega$$

where θ is the angle through which the incident particle is scattered and b is the distance of closest approach. If Ze is the charge on the nucleus, $Z'e$ and v the charge and velocity respectively of the incident particle and μ is the reduced mass, then

$$b = \frac{Z Z' e^2}{\mu v^2}.$$

Rydberg *See* **Appendix**

Rydberg Constant *See* **Ritz Combination Principle** and **Bohr Theory of Spectral Emission**

Rydberg-Schuster Law
The difference of the wave-numbers of the limit of the principal series and the common limit of the diffuse and sharp series in alkali spectra is equal to the wave-number of the first line of the principal series.

S

Sabatier-Senderens Reduction
When a mixture of carbon monoxide or carbon dioxide and hydrogen is passed over finely divided nickel at 300°C methane is formed. Many other organic compounds may be reduced in a like manner and all these reductions are known by this name.

Sabin *See* Appendix

Sabine Formula
In acoustics a formula giving the reverberation time of an enclosure

$$T = \frac{V}{20 \sum\limits_{r=1}^{n} a_r S}$$

where
T = reverberation time in seconds,
V = Volume of enclosure in cu. ft,
$a_1, a_2, a_3, \ldots$ = absorption coefficient of areas $S_1, S_2, S_3, \ldots$

Sachse-Mohr Theory

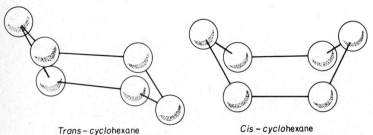

Trans – cyclohexane   Cis – cyclohexane

Sachse-Mohr Theory

The stability of carbocyclic rings appears to increase up to five- and six-membered rings and then to remain effectively constant. This fact can be explained if these rings are assumed to be non-planar thus

272

assuming that rings with six or more carbon rings are puckered. Thus according to Sachse *cyclo*hexane exists in two forms as shown in the figure.

These two forms have never been isolated and, indeed, calculation shows that they would differ in energy content by only 5 kcal/mole and ready conversion of one form to the other would produce apparently a planar molecule. Mohr suggested, however, that if two *cyclo*hexane rings are fused together as in decalin the two forms would be stable enough to exist independently of one another. Hückel managed later to separate the two forms. This theory, therefore, does not deny the existence of large rings and predicts that if they could be prepared they would be stable. Ruzieka's work from 1926 onward has shown this to be the case.

Sackur-Tetrode Equation
This equation allows the translational entropy, a quantity identical with total molar entropy, to be calculated. If S_{tr} is the total molar entropy of a monatomic gas and m is the mass of a particle inhabiting a box of volume V then

$$S_{tr} = R \left\{ \ln \left(\frac{(2\pi mkT)^{3/2}}{h^3 N} \right) V + \frac{5}{2} \right\}$$

Saha's Equation
An equation giving the degree of thermal ionization ϵ_e of a monatomic gas

$$\log \frac{\epsilon_e^2}{1 - \epsilon_e^2} P(\text{atm}) = -5040 \frac{E}{T} + \frac{5}{2} \log T + \log \frac{\omega_i \omega_e}{\omega_a} - 6 \cdot 491$$

where ω_i, ω_e and ω_a are constants that refer respectively to ion, electron and atom.

P is the pressure in atmospheres and E is the ionization potential in volts.

Saint Venant Plasticity
Ideal plasticity in which plastic deformation proceeds at constant yield stress.

273

Saint Venant's Principle
If one applies to a small part of the surface of a body a set of forces which are statically equivalent to zero, then this system of forces will not noticeably affect parts of the body lying away from the above region.

Sakaguchi's Arginine Reaction
To about 3 c.c. of the solution add 1 c.c. of 5% sodium hydroxide and 2 drops of a 1% solution of a-naphthol in alcohol. Add a single drop of 10% sodium hypochlorite and shake. A bright red colour develops if arginine is present. Ammonia, however, interferes.

Salkowski's Test
A common test for cholesterol which consists in treating the sterol with chloroform and concentrated sulphuric acid to give a bluish red or purple colour.

Sandmeyer Reaction
When a diazonium salt is treated with a solution of a cuprous halide in the corresponding halogen acid the diazo group is replaced by a halogen.

Sarrus' Rule
A rule for determining signs of terms in the expansion of a determinant of order three. Write down the determinant and then repeat the first and second columns, whereby the signs of the terms in the expansion are given by the scheme:

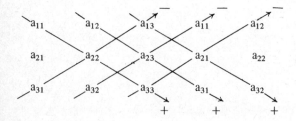

Schardinger Reaction
The reduction of methylene blue by an aldehyde occurs more quickly in the presence of fresh milk than boiled milk. Ball considers that the enzyme responsible for this catalysis is xanthine oxidase, sometimes known as Schardinger's enzyme.

Schemann Reaction

The controlled thermal decomposition of dry aromatic diazonium fluoborates to give an aromatic fluoride, boron trifluoride and nitrogen.

Schiff Bases

Aniline can be condensed with aromatic aldehydes to form anils (Schiff's Bases).

These bases are readily hydrolysed to the free amine and may therefore be used to protect an amine group during other processes, e.g. nitration. On reduction the anils give secondary amines and they have been used to prepare such compounds

Schiff Test for Aldehydes

Rosaniline is dissolved in water and sulphur dioxide is passed through the solution until it is decolourized. Aldehydes cause the magenta colour to return to this solution.

Schläfli's Integral

$$P_n(x) = \frac{1}{2\pi j} \oint \frac{(t^2 - 1)^n \, dt}{2^n \, (t - x)^{n+1}}$$

where $P_n(x)$ is the **Legendre Coefficient** of order n. Integration is anticlockwise around a contour C encircling z in the complex plane.

Schlomilch's Expansion

Any function f(x) which has a continuous differential coefficient in

275

the range $0 < x < \pi$ may be expressed as a series of **Bessel Functions**.

$$f(x) = a_0 + \sum_{m=1} a_m J_0 \, (mx)$$

where

$$a_0 = f(0) + \frac{1}{\pi} \int_0^\pi u \int_0^{\pi/2} f'(u \sin \theta) \, d\theta \, du$$

$$a_m = \frac{2}{\pi} \int_0^\pi \cos mu \int_0^{\pi/2} f'(u \sin \theta) \, d\theta \, du$$

Schlomilch's Infinite Product *See* **Weierstrass' Infinite Product**

Schmid's Law of Critical Stress Shear
Slip in a material takes place along a given slip plane and direction when the shear stress acting along them reaches a critical value.

Schmidt Number
A dimensionless number used in diffusion and given by $(\eta/\rho) \, D_{AB}$ where D_{AB} is the mass diffusivity of species A in species B, η is the dynamic viscosity and ρ the density.

Schmidt Reaction
Carbonyl compounds react with hydrazoic acid in the presence of concentrated sulphuric acid. The aldehydes are converted into cyanides and formyl derivatives.

$$RCHO + HN_3 \xrightarrow[-N_2]{H_2SO_4} RCN + RNH.CHO$$

Schmitt Synthesis *See* **Kolbe Synthesis**

Schoenflies Crystallographic Notation
Schoenflies treated the crystallographic symmetries as mathematical 'groups' of operations. An operation and its powers are called a *cyclic group*, thus a fourfold axis is represented by the operations:
   $1, \pi/2, \pi, 3\pi/2$, i.e. powers 0, 1, 2, 3 of the fundamental rotation $\pi/2$. The symbol for a cyclical group for a rotation axis of order n is C_n. Other operations are defined by various other symbols.
   Dihedral groups (a set of twofold axes at right angles to a major axis of order n) by D_n.

Octahedral group (several axes orientated along the rational directions of a cube) by O.

Tetrahedral group (the symmetries are both those of the octahedron and tetrahedron) by T.

The eleven axial symmetries may be represented by the symbols $C_1, C_2, C_3, C_4, C_6, D_2, D_3, D_4, D_6, O$ and T. Other groups can be formed by adding inversion centres and reflection planes. Inversion centres are represented by subscript i and reflection planes by subscript v, h and d dependent on whether the plane is vertical, horizontal or diagonal.

The names International and Schoenflies symbols are given in the following tables (after Buerger).

The symbol $S_4 \equiv \bar{4}$ is the only group which cannot be described by adding inversion centres or minor planes to the original eleven symbols.

In the International Symbols (**Hermann-Mauguin Symbols**), 1, 2, 3, 4 and 6 represent 1-, 2-, 3-, 4- and 6-fold rotation axes, m represents a mirror plane of symmetry and $\bar{1}, \bar{2}, \bar{3}, \bar{4}$ and $\bar{6}$ are axes of rotation-inversion.

Crystal System	Schoenflies Notation	International Symbol	Crystal System	Schoenflies Notation	International Symbol
Triclinic	C_1	1	Hexagonal	C_3	3
	C_i	$\bar{1}$		D_3	32
				C_{3i}	$\bar{3}$
Monoclinic	C_2	2		C_{3v}	3m
	C_{1h}	m		D_{3d}	$\bar{3}\frac{2}{m}$
	C_{2h}	$\frac{2}{m}$			
				C_{3h}	$\bar{6}$
				D_{3h}	$\bar{6}2\,m$
Orthorhombic	D_2	222		C_6	6
	C_{2v}	2mm		D_6	622
	D_{2h}	$\frac{2}{m}\frac{2}{m}\frac{2}{m}$		C_{6v}	6mm
				C_{6h}	$\frac{6}{m}$
Tetragonal	C_4	4			
	D_4	422		D_{6h}	$\frac{6}{m}\frac{2}{m}\frac{2}{m}$
	S_4	$\bar{4}$			
	C_{4h}	$\frac{4}{m}$	Cubic	T	23
				O	432
	C_{4v}	4mm		T_h	$\frac{2}{m}\bar{3}$ (m3)
	D_{2d}	$\bar{4}2m$		T_d	$\bar{4}3m$
	D_{4h}	$\frac{4}{m}\frac{2}{m}\frac{2}{m}$		O_h	$\frac{4}{m}\bar{3}\frac{2}{m}$, m 3 m

Schonherr Process

A process for the direct fixation of atmospheric nitrogen using an arc furnace.

Schotten-Baumann Reaction

Compounds containing an active hydrogen atom can be benzoylated by means of benzoyl chloride in the presence of dilute caustic soda.

$$C_6H_5NH_2 + ClOC.C_6H_5 \rightarrow C_6H_5NH.OC.C_6H_5 + HCl$$

Schottky Defect *See* Frenkel Defect

Schottky Anomaly

In a system having two (or more) energy levels such that excitation can occur between these energy levels, the variation of specific heat with temperature shows a maximum corresponding to the temperature region where there is strongest excitation to the higher energy level. The behaviour, referred to as the Schottky anomaly, has been noted particularly in paramagnetic salts.

Schrödinger Wave Equation

This equation is fundamental to wave mechanics and has been used to solve a number of problems relating to atomic and molecular structure.

$$\frac{\partial^2 \psi}{\partial x^2} + \frac{\partial^2 \psi}{\partial y^2} + \frac{\partial^2 \psi}{\partial z^2} - \frac{8\pi^2 m}{h^2} (E - V) \psi = 0$$

where ψ is the amplitude function for three dimensions, E is the total energy and V is the potential energy of the particle mass m.

Schulz and Sing Formula

A number of empirical relationships have been proposed to relate the concentration and viscosity number of a viscous solution.

Schulz and Sing	$[\eta]c = \eta_{sp}/(1 + k\eta_{sp})$
Huggins	$\eta_{sp} = [\eta]c + k.[\eta]^2c^2$
Makin	$\eta_{sp} = [\eta]c \exp k[\eta]c$

where $[\eta]$ and k are constants. η_{sp} is the specific viscosity and c is the concentration.

Schulze-Hardy Rule

The ion of an electrolytic which is effective in bringing about the coagulation of a sol is of opposite charge to that on the colloidal particles and the coagulating power increases considerably with increasing valency of the ion.

Schur's Lemma

A reducible representation of a group is one in which all the matrices A, B, C, . . . can be transformed by a unitary matrix M such that $M^{-1}AM$ etc. have the same block diagonal form; in this all the elements are zero except for those in a set of squares along the main (leading) diagonal. Then, **Schur's Lemma** states that any matrix which commutes with all the matrices of an irreducible representation of a group is a constant matrix, i.e. a multiple of the unit matrix.

Schwarz-Christoffel Transformation

A conformal transformation of the inside of a polygon into the upper half of a complex plane.

Schwarz Principle of Reflection

If $f(x)$ is analytic within a region D intersected by the real axis and is real on the real axis, then, for conjugate values of z, conjugate values of f are obtained, i.e.

$$f(z^*) = f^*(z)$$

where * denotes the conjugate quantity.

Schwarz's Inequality

If f and g are any two functions then

$$\int |f|^2 \, d\tau \int |g|^2 \, d\tau \geqslant |\int f^* g \, d\tau|^2$$

where f^* is the complex conjugate of f, and integration is over all space. The equality only applies if one function is a scalar multiple of the other; i.e. if $g = \lambda f$ where λ is a constant.

The equivalent theorem in three-dimensional vector space is

$$(\mathbf{v}_1 \cdot \mathbf{v}_1)(\mathbf{v}_2 \cdot \mathbf{v}_2) \geqslant (\mathbf{v}_1 \cdot \mathbf{v}_2)^2.$$

(*See also* **Cauchy's Inequality**.)

Schweitzer's Reagent

An ammoniacal solution of copper hydroxide and a solvent for cellulose.

Secchi's Classification
Father Secchi (1818-1878) divided about 4000 stars into four groups. This represented the earliest classification of stars into spectral types.

Seebeck Effect
If a closed electrical circuit is composed of two dissimilar metals, a current flows round the circuit if one of the junctions is maintained at a different temperature from the other. The direction and magnitude of the current depend on the nature of the metals as well as on the temperatures of the junctions. An absolute measure of the effect (often called the thermoelectric power) for one metal can be obtained if the other metal is superconducting.

Seidal Aberrations
The primary aberrations found in lens systems. Deviations of light rays from the paths prescribed by **Gauss' Optics Formulae** can be expressed in terms of five sums called **Seidal sums** which must all be zero for no monochromatic aberration, coma, astigmatism, field curvature and distortion, and it is not possible to eliminate all these aberrations at one time.

Seiffert's Spherical Spiral
A curve described on a sphere. If a sphere is taken with centre at the origin and radius unity, and if the cylindrical polar coordinates of any point on it be (ρ, ϕ, z) so that the arc of a curve traced on the sphere be given by

$$(ds)^2 = \rho^2(d\phi)^2 + (1 - \rho^2)^{-1} (d\rho)^2.$$

Seiffert's spiral is defined by

$$\phi = ks$$

where s is the arc measured from the pole of the sphere, and k is a positive constant, less than unity.

Seignette Electricity
Now known as ferroelectricity. The phenomenon was first observed on Seignette or Rochelle Salt, which was first made as a laxative by the pharmacist Seignette (1672) living in La Rochelle.

Seliwanoff's Test
A specific test for ketoses which, when treated with hydrochloric acid and resorcinol, give rise to a red coloration.

Sellmeyer Formula *See* Ketteler-Helmholtz Formula

Serber Potential
A distribution of potential applicable to neutron-proton scattering at low energies arising from the presence of **Majorana** and **Wigner Forces**.

Serret-Frenet Formulae
If $\delta\theta$ is the small angle between tangents at neighbouring points P and Q on a curve, $\mathbf{t}$ is the unit vector along the tangent to the curve at P, $\mathbf{n}$ the unit vector in the principal normal plane given by $\mathbf{n} = d\mathbf{t}/d\theta$ for $\delta\theta \to 0$, and $\mathbf{b}$ is the unit vector orthogonal to $\mathbf{t}$ and $\mathbf{n}$, and making a right-handed triad of unit vectors in that order, then

$$\mathbf{t}' = \kappa\mathbf{n}$$
$$\mathbf{b}' = -\tau\mathbf{n}$$
$$\text{and } \mathbf{n}' = -\kappa\mathbf{t} + \tau\mathbf{b}$$

where κ is the curvature and τ is the angle turned through per unit length of curve at the point P on the curve.

Shannon's Sampling Theorem
A function $x(t)$ having a power spectrum which is restricted to a finite frequency band is uniquely determined by a discrete set of sample values. Thus, if $x(t)$ is restricted to the frequency interval $-B \leqslant \nu \leqslant B$, then

$$x(t) = \sum_{k=-\infty}^{\infty} x\left(\frac{k}{2B}\right) \frac{\sin(2Bt - k)}{\pi(2Bt - k)} \qquad (-\infty < t < \infty).$$

Sheppard's Correction
A correction factor used in statistics when calculating standard deviation σ for data which is grouped into classes of width c.
If $\mu^2 = \Sigma f_s (x_s - m)^2/\Sigma f_s$ (see **Charlier's Checks** for the notation) and the class width is c about x_s, then the value of μ^2 and hence σ^2 should be reduced by $c^2/12$ to allow for the class width.

Shockley Partial Dislocation

The strain energy of a dislocation is proportional to b^2 where b is the magnitude of **Burger's Vector**. Because of the consequent reduction in strain energy, some dislocations split into two partials having smaller **b** than the complete dislocation. If the partial has **b** lying in the fault plane it is a Shockley partial dislocation (also called a 'glissile' dislocation because it is able to glide); if **b** is not parallel to the fault plane it is a **Frank Partial Dislocation** (also called a 'sissile' dislocation because it can only diffuse and not glide).

Shore Scleroscope Hardness

The height of rebound of a diamond-pointed hammer falling under its own weight on the object whose hardness is to be measured. The hardness measurement is based on an empirical scale wherein the average hardness of martensitic high-carbon steel is equal to 100.

Shubnikov-de Haas Effect

Oscillations in the magnetoresistance or **Hall Effect** in a metal or semiconductor as a function of a strong magnetic field, arising from the quantization of the energy of the electrons. (*See also the* **de Haas-van Alphen Effect.**)

Shubnikov Groups

Alternatively called magnetic groups or black and white groups, they are point groups of crystals in which a magnetic moment is present. In two dimensions the point group 4 *m m* produces two magnetic point groups 4′ *m m*′ and 4 *m*′ *m*′ where black and white shading is equivalent to magnetic moments in opposite directions;

4 mm   4′ mm′   4 m′m′

Shubnikov Groups

In three dimensions, the 32 ordinary groups (for their tabulation see **Schoenflies Crystallographic Notation**) become 58 black and white point groups. Similarly, 230 ordinary space groups become 1191 black and white space groups.

Siegbahn X Unit *See* **Appendix**

Siemens *See* **Appendix**

Silsbee Effect

Kamerlingh Onnes (1913) found that if an electric current is passed down a superconductor, then when it reaches a critical value, it destroys the superconductivity. Silsbee (1916) showed that the transition back to the normal state depends on the magnetic field associated with the current. The critical value of the magnetic field required to destroy the superconductivity is a function of temperature.

Simpson's Line

The feet of the perpendiculars drawn from any point on the circumcircle of a triangle to the three sides lie in a straight line, known as Simpson's Line.

Simpson's Rule

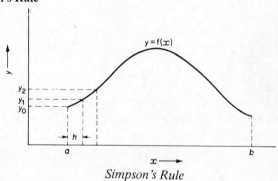

Simpson's Rule

A method of computing definite integrals. Let

$$A = \int_a^b f(x)\,dx$$

be interpreted as the area between the x axis and the curve $y = f(x)$ bounded by the ordinates $x = a$ and $x = b$. Divide the interval a, b into n equal parts and call $(b - a)/n = h$ and $y_0, y_1, \ldots$ the values of $f(x)$ at $a, a + h, \ldots$; then

$$A = \frac{h}{3}\left(y_0 + 4y_1 + 2y_2 + 4y_3 + 2y_4 + \ldots + 4y_{n-1} + y_n\right).$$

283

Skraup Synthesis

Quinoline may be prepared by heating together aniline, nitrobenzene, glycerol, concentrated sulphuric acid and ferrous sulphate.

Nitrobenzene acts as an oxidizing agent and ferrous sulphate makes the reaction less violent. Arsenic acid may be used instead of nitrobenzene. The reaction mechanism is obscure but it has been assumed that it may be represented as

Snell's Law

The relative refractive index is given by the ratio of the sine of the angle of incidence to the sine of the angle of refraction for a ray passing from one isotropic medium into another isotropic medium and in which the velocity of propagation decreases from the first to the second medium.

Snoek's Law

Alloys with an integral number of Bohr magnetons per atom show very little, if any, magnetostriction.

Soddy-Fajans Displacement Law

The emission of an α-particle from an atom produces a displacement to the left by two places in the periodic table and the emission of a β-particle produces a displacement to the right by one place; i.e. the atomic numbers are reduced by 2 and increased by 1 respectively. This law is the basis of the radioactive series.

Solvay Process (for the Preparation of Sodium Carbonate)
The Solvay or ammonia-soda process depends upon the comparatively small solubility of sodium bicarbonate. The reaction occurs in three stages.

(1) Brine is saturated with ammonia gas and carbon dioxide. The solubility product of sodium bicarbonate is low and so it is precipitated.

(2) The sodium bicarbonate is filtered off and calcined.

$$2NaHCO_3 \rightarrow Na_2CO_3 + H_2O + CO_2$$

(3) The ammonia is regenerated from the mother liquor by the action of lime.

Sommelet Reaction
Aryl aldehydes may be prepared by the reaction of benzyl halides and hexamethylene tetramine by heating the reactants together in alcohol (see **Delepine Reaction**).

$$ArCH_2Cl + (CH_2)_6N_4 \xrightarrow{H_2O} ArCH_2NH_2 \xrightarrow{HCHO} ArCH_2N{=}CH_2$$

$$ArCHO + H_2NCH_3 \xleftarrow{H_2O} ArCH{=}NCH_3$$

Sommerfeld Fine Structure Constant *See* **Appendix**

Sommerfeld's Theory (of the Thermal Conductivity of Metals)
A theory of thermal conductivity depending upon the free electron theory of metals. It was assumed that the electrons in the metal behave as a degenerate gas at ordinary temperatures and the distribution of energy among the electrons is not given by **Maxwell's Law** but by **Fermi-Dirac Statistics**.

Sørensen pH Scale
The hydrogen ion concentration of a solution can vary from 1 gm atom per litre for a strong acid to 10^{-14} gm atoms per litre for a strong alkali. Sørensen suggested a more convenient way of expressing these quantities to avoid using negative powers of ten. He suggested that it should be expressed as pH, where

$$pH = -\log a_{H+}$$

where a_{H+} is the activity of the hydrogen ion which for dilute solutions may be replaced by the concentration.

285

Sørensen Titration

Acids such as glycine because of their amphoteric nature cannot be titrated directly with alkali. When formalin solution is added a methylene amino-acid is formed.

$$H_2N.CH_2COOH + HCHO \rightarrow CH_2.N.CH_2.COOH + H_2O$$

This is a strong acid in which the amine group is protected and hence may be titrated with sodium hydroxide.

Soret Effect

If a thermal gradient is applied to a mixture or solution, a migration of atoms occurs by thermal diffusion so that a concentration gradient is set up. This is known as the Soret effect. The converse process whereby concentration gradients produce non-uniformity of temperature is termed the **Dufour effect**.

Spiegler's Test (for Proteins in Urine)

Render the urine faintly acid with acetic acid and filter if necessary. Place 3 c.c. of Spiegler's Reagent in a narrow test tube. Float 3 c.c. of urine on the surface. A white ring formed at the junction of the two liquids indicates proteins. This reaction is also given by proteoses and peptoses. This test is very delicate and is too sensitive for ordinary clinical work as a large number of apparently normal urines give a positive reaction. Spiegler's reagent is:

Mercuric chloride	40 gm
Tartaric acid	20 gm
Glycerol	100 gm
Sodium chloride	50 gm
Distilled water	1000 c.c.

Stahl's Phlogiston Theory

All combustible bodies contain an elemental substance, phlogiston, which could be driven out by heat. The ash or residue left after combustion represented the original matter less its phlogiston.

Standt's Theorem

Every Bernoulli number B_{2n} is equal to an integer diminished by the sum of the reciprocals of all and only those prime numbers which, when diminished by unity, are divisors of $2n$.

Stanton's Number
The inverse of **Prandtl's Number**.

Stark Effect
Stark, in 1913, observed an effect, analogous to the transverse **Zeeman Effect**, on the hydrogen spectrum when hydrogen atoms were subjected, while radiating, to an electric field. He also observed a longitudinal effect.

In the transverse Stark effect, components exist polarized both parallel and at right angles to the field. The position and polarization of these components have been accounted for by a theory which constitutes, perhaps, one of the greatest successes of the quantum theory.

Stark-Einstein Law
Each molecule taking part in a chemical reaction induced by exposure to light absorbs one quantum of radiation causing the reaction.

Stefan's Constant *See* **Appendix**

Stefan's Law *See* **Boltzmann Law of Radiation**

Steffens Process
A method of producing sucrose from molasses by diluting to a solution containing 7% sugar and adding lime which precipitates the slightly soluble tricalcium saccharate. This compound is filtered off and added to raw sugar which is then heated to produce a sugar solution and a scum which is removed.

Steiner's Theorem
The moment of inertia of a body depends not only on the total mass and its distribution but also upon the axis of rotation. In particular it changes when the axis is displaced parallel to itself. Let I_c be the moment of inertia of a body of total mass M about an axis through its centre of gravity, let I_a be the moment of inertia about another axis parallel to the first and at a distance a from it. Then Steiner's Theorem gives a simple relation between these quantities.

$$I_a = I_c + Ma^2$$

Steinmetz's Law
The energy dissipated per unit volume per cycle during a magnetic hysteresis loop is given by

$$W = \eta\, B_{max}^{1 \cdot 6}$$

where B_{max} is the maximum magnetic induction obtained during the cycle and η is Steinmetz's coefficient which is a constant for a given material. The law holds only within a limited range of induction and in certain materials there is a slight deviation from the power of $1 \cdot 6$.

Stephen Reaction
Aromatic nitriles can be reduced by stannous chloride and hydrogen chloride to imino-chlorostannates which hydrolyse easily to give the aldehyde.

$$ArCN + SnCl_2 + 4HCl \rightarrow [ArCH{=}NH_2]\,[HSnCl_6]$$
$$\downarrow H_2O$$
$$ArCHO + (NH_4)HSnCl_6$$

Stern-Gerlach Effect
If a beam of randomly orientated atoms is passed through an inhomogeneous field, then the beam is split into a number of beams, this number depending on the kind of atom and its state. The effect shows the quantization of magnetic moment since classically the beam should be deflected into a vertical line.

Stewart and Kirchhoff's Law
The emissive power divided by the absorption coefficient for any substance depends only on the frequency and plane of polarization of the radiation and the temperature and is independent of the nature of the substance.

Stieltjes Integral
Let an interval (a, b) be divided into sub-intervals in the following manner

$$a = x_0 < x_1 < x_2 < \ldots < x_n = b$$

and let $\triangle$ equal the largest of these sub-intervals. Then the Stieltjes integral of $f(x)$ with respect to $a(x)$ from a to b is

$$\int_a^b f(x)\, da(x) = \lim_{\triangle \to 0} \sum_{k=1}^{n} f(\zeta_k)[a(x_k) - a(x_{k-1})]$$

where $x_{k-1} \leqslant \zeta_k \leqslant x_k$.

Stiles-Crawford Effect

Light entering the eye near the centre of the pupil is usually more effective in producing a visual response than light entering near the perimeter. The effect is attributed to a variation of response of the receptors of the retina with direction of the light.

Stirling's Formula

Valid for large values of n.

$$\ln n! = (n + \tfrac{1}{2}) \ln n - n + \tfrac{1}{2} \ln 2\pi + \frac{1}{12n} - \text{terms of the order of } n^{-3}$$

Stirling's Series (for the Gamma Function)

The gamma function (*see also* **Euler's Definition of the Gamma Function**) can be expressed as

$$\Gamma(z) = e^{-z}\, z^{z-\frac{1}{2}}\, \sqrt{(2\pi)}\, \left(1 + \frac{1}{12z} + \frac{1}{288z^2} - \frac{139}{51840z^3}\right.$$

$$\left. + O(z^{-4})\right)$$

($|\arg z| < \pi$). For z real and positive, the absolute value of the error is less than the last term.

Stobbe Condensation

The condensation between a succinic ester and carbonyl compound in the presence of sodium ethoxide. The yield may be increased by using potassium t-butoxide in t-butanol as the condensing agent.

$$\begin{array}{c}
\diagdown \\
\diagup
\end{array}C{=}O + CH_2COOC_2H_5 \qquad \begin{array}{c}\diagdown\\\diagup\end{array}C{=}C.CH_2COOH$$

$$\begin{array}{c}\quad | \\ CH_2COOC_2H_5\end{array} \xrightarrow[\text{t-butanol}]{\text{t-butoxide}} \begin{array}{c} | \qquad\qquad + H_2O \\ COOC_2H_5\end{array}$$

The condensation product may be refluxed with hydrobromic acid in glacial acetic acid and the γ lactone formed, catalytically reduced via the sodium salt to the acid.

Stockbarger Method *See* Bridgman-Stockbarger Method

Stockmayer Potential

The total potential energy Φ of two polar (ionic) molecules is of the form

$$\Phi = \frac{B}{r^{12}} - \frac{C}{r^6} - \frac{\mu^2}{r^3} \left\{ 2 \cos \theta_1 \cos \theta_2 - \sin \theta_1 \sin \theta_2 \cos (\phi_2 - \phi_1) \right\}.$$

The first two terms are as for the **Leonard-Jones Potential** whereas the third term represents the interaction between two ideal dipoles described by angles θ_1 and ϕ_1 and θ_2 and ϕ_2; μ is the dipole moment of a single molecule and r the separation.

Stokes *See* Appendix

Stokes' Law

If the resistance to motion is R in a fluid of viscosity η for a particle of radius r moving at a velocity v then

$$R = 6\pi\eta rv.$$

Stokes' Law of Radiation

It was supposed by G. Stokes that in fluorescence only radiations of wavelength longer than that of the exciting light could be emitted. In line spectra these are known as **Stokes' Lines**.

Emitted lines of shorter wavelength than the incident light have also been observed (for example, in the Raman Spectra of benzene), and are called **Anti-Stokes' Lines**. The difference in wavelength between the emitted and exciting light is called **Stokes Shift**.

Stokes' Phenomenon

An apparent discontinuity in the value of an asymptotic expansion of a function as its argument changes its phase.

Stokes' Theorem
The surface integral of the curl of a vector over any surface is equal to the line integral of the vector round the boundary of the surface.

$$\iint \text{curl } \mathbf{A} \cdot d\mathbf{S} = \int \mathbf{A} \cdot ds$$

Stormer Unit *See* **Appendix**

Strecker Degradation
An aromatic compound which contains a carbonyl group in conjunction with another carbonyl group or nitrogroup will react with an a-amino acid to give the aldehyde and carbon dioxide. Measurement of the volume of carbon dioxide evolved may be used to estimate quantitatively the a-amino acid.

Strecker Reaction
When an alkyl halide is heated with sodium sulphite the sodium salt of the sulphonic acid is produced.

$$RX + Na_2SO_3 \rightarrow RSO_3Na + NaX$$

Strouhal Number
A dimensionless number concerning non-steady flow of a viscous fluid past a solid body. If v is the velocity of flow of the fluid, τ a characteristic time interval for the flow (related to the rate of change) and l a characteristic length for the body, then the Strouhal number is given by $S = v\tau/l$. See also **Reynolds Number.**

Struve Functions
The Struve function of the mth order is

$$S_m(z) = \frac{2(z/2)^m}{\Gamma(m + \frac{1}{2})\,\Gamma(\frac{1}{2})} \int_0^{\pi/2} \sin(z \cos u) \sin^{2m} u \cdot du.$$

Sturm-Liouville Equation

$$\frac{d}{dz}\left(p(z)\,\frac{dy}{dz}\right) + \left(q(z) + \lambda\, r(z)\right)y = 0$$

where λ is the eigenvalue.

Sturm's Functions

Let $f(x)$ be a polynomial and $f'(x)$ its derivative. When $f(x)$ is divided by $f'(x)$, let the remainder be $-f_2(x)$. Similarly, if $f'(x)$ is divided by $f_2(x)$, the remainder is called $-f_3(x)$. Thus,

$$f(x) = g_1(x) f'(x) - f_2(x)$$
$$f_1(x) = g_2(x) f_2(x) - f_3(x) \qquad (f_1(x) = f'(x))$$
$$\ldots \qquad \ldots \qquad \ldots \qquad \ldots$$
$$f_{n-2}(x) = g_{n-1}(x) f_{n-1}(x) - f_n(x).$$

The degrees of the functions $f(x)$, $f'(x)$, $f_2(x) \ldots f_n(x)$, Sturm's functions, steadily decrease, and the final remainder, $-f_n(x)$, is a constant which is only zero if $f(x) = 0$ has a multiple root.

Sturm's Theorem

If $f(x)$ is a polynomial and a and b are any real numbers ($a < b$), the number of distinct roots of $f(x) = 0$ which lie between a and b (any multiple roots which may exist being counted only once) is equal to the excess of the number of changes of sign in the sequence of **Sturm's Functions**

$$f(x), f'(x), f_2(x), \ldots, f_n(x)$$

where $x = a$, over the number of changes of sign in the sequence when $x = b$.

Suida Process

Pyroligneous liquor from the destructive distillation of wood contains 4–10% of acetic acid. This acid may be recovered by extraction from the vapour with tar oil.

Sullivan's Test (for Cystine)

To 2 c.c. of a dilute solution (0·1% of 0·1N HCl) add 3 or 4 drops of a 2% aqeous solution of 1 : 2 naphthoquinone-4-sodium sulphonate and 2 c.c. of 20% sodium sulphite in 0·5N NaOH. Add a few drops of 5% sodium cyanide and allow to stand for 10 minutes. A brown red coloration is obtained. Now add 1 c.c. of 2% sodium hydrosulphite in 0·5N NaOH. The colour is intensified and changes to pure red. Many amino acids give the first coloration but are reduced to colourless compounds by the hydrosulphite.

Sutherland's Formula
The viscosity η of a gas at temperature T is given by

$$\eta = \eta_0 \left(\frac{T}{273\cdot1}\right)^{3/2} \frac{C + 273\cdot1}{C + T}$$

where η_0 is the viscosity at $0°C$ and C is a constant for a given gas.

Svedberg *See* Appendix

Swarts' Reaction
Alkyl fluorides may be prepared by heating organic halides with inorganic fluorides such as AsF_3, SbF_3, AgF or Hg_2F_2.

$$C_2H_5Cl + AgF \rightarrow C_2H_5F + AgCl$$

Sylvester's Theorem
In matrix theory: If the n latent roots, (eigenvalues), $\lambda_1, \lambda_2, \ldots, \lambda_n$ of the square matrix A, are all distinct and $P(A)$ is any polynomial of A,

$$P(A) = \sum_{r=1}^{n} P(\lambda_r)\, Z_r$$

where the matrix Z_r is

$$Z_r = \frac{\prod_{s \neq r} (A - \lambda_s U)}{\prod_{s \neq r} (\lambda_r - \lambda_s)}$$

and U is the unit matrix.

Szent-Györgi Cycle
Szent-Györgi has advanced the hypothesis that certain carbon acids, fumaric, malic and oxalacetic acids and more particularly succinic acid act as 'carriers' in biological oxidations. He was led to this view by the observation that most tissues oxidize succinic acid to fumaric acid more rapidly than any other substance known and that respiration was definitely poisoned by the addition of minute quantities of malonic acid which is a specific inhibitor for the above reaction.

T

Tafel Equation

$$\eta = a + b \log I$$

in electrode kinetics, where η is the polarization or overpotential of the electrode in the electrolyte, I is the current flowing between electrode and electrolyte and

$$a = \mp (2 \cdot 303 RT/aF) \log I_0$$
$$b = 2 \cdot 303 RT/aF$$

where a is the fraction of η assisting the dissolution of the electrode and I_0 is the exchange current or the current per unit area of electrode at the reversible potential when $\eta = 0$.

When η is plotted as ordinate against $\log I$ the curve is known as the **Tafel line**.

Talbot *See* **Appendix**

Talbot's Bands

If a thin glass plate is placed across half the aperture of a prism so that it is in front of the *thinner* half of the prism, or if the plate is placed similarly across half the width of a grating, then a series of dark bands called Talbot bands appear in the spectrum if white light is incident on the prism or grating. These bands appear because of the path difference introduced between light passing through the two sides of the prism or grating. There is an optimum thickness of glass plate to produce maximum visibility and if the thickness is above a critical value the bands disappear because retardation of the light waves becomes too great for interference.

Talbot's Law

A light flashing at a frequency greater than approximately 10 hertz appears steady to the eye due to the persistence of vision. By Talbot's Law the light has an apparent intensity I given by

$$I = I_0 \left(\frac{t}{t_0} \right)$$

where I_0 is the actual intensity of the light source for exposure time t during total time t_0. See also **Blondel-Rey Law**.

Tammann Temperature
An approximate minimum temperature at which two solids will react together at an appreciable rate.

Tate's Law
A drop falls from a narrow tube when the weight of the drop is equal to the surface tension which holds it to the liquid: $mg = 2\pi\gamma r$, where m is the mass of a drop of radius r and γ is the surface tension.

Taylor-Orowan Dislocation
An alternative name for an edge dislocation in a crystal. This can be considered as the presence in the crystal of an extra half plane of atoms; the coordination of the extra atoms in the dislocation is less than that for the ideal lattice. See also **Burgers' Circuit**.

Taylor's Theorem

$$f(x + h) = f(x) + \frac{h}{1!} f'(x) + \frac{h^2}{2!} f''(x) + \ldots + \frac{h^n}{n!} f^n(x) + R_n$$

R_n, the remainder, is given as

$$R_n = f^{(n+1)}(x + \theta h) \frac{h^{n+1}}{(n+1)!}; \quad (0 < \theta < 1) \quad \text{Lagrange's form.}$$

$$R_n = f^{(n+1)}(x + \theta h) \frac{h^{n+1}(1 - \theta)^n}{n!}; \quad (0 < \theta < 1) \quad \text{Cauchy's form.}$$

Taylor's theorem for functions of many variables is:

$$f(x_1 + h_1, x_2 + h_2, \ldots)$$

$$= f(x_1, x_2, \ldots) + h_1 \frac{\partial f}{\partial x_1} + h_2 \frac{\partial f}{\partial x_2} + \ldots$$

$$+ \frac{h_1{}^2}{2!} \frac{\partial^2 f}{\partial x_1{}^2} + \frac{2}{2!} h_1 h_2 \frac{\partial^2 f}{\partial x_1 \partial x_2} + \frac{h_2{}^2}{2!} \frac{\partial^2 f}{\partial x_2{}^2} + \ldots$$

Tesla *See* **Appendix**

Thénard's Blue Reaction
A reaction for aluminium, based on the fact that when cobalt oxide is ignited with aluminium oxide, a double blue oxide of spinel type is formed.

Thévénin's Theorem

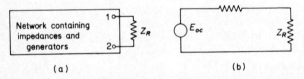

(a)   (b)

Thévénin's Theorem

The current in any impedance Z_R, connected to two terminals of a network, is the same as if Z_R were connected to a simple generator, whose generated voltage is the open-circuited voltage at the terminals in question and whose impedance is the impedance of the network looking back from the terminals, with all generators replaced by impedances equal to the internal impedances of these generators.

It has been stated that this theorem was originally enunciated by Helmholtz and should, more correctly, be called Helmholtz's theorem.

Thiele Benzene Formula *See* **Kekulé Benzene Formula**

Thiele Theory of Partial Valencies
When halogens add on to compounds containing conjugated double

bonds the expected 1 : 2 adduct is formed but abnormal 1 : 4 addition occurs as well.

$$CH_2{=}CH{-}CH{=}CH_2$$

$$\xrightarrow{Br_2} CH_2Br{-}CHBr{-}CH{=}CH_2 + CH_2Br{-}CH{=}CH{-}CH_2Br$$

These two reactions always occur together and the relative proportions of the two adducts depend upon such factors as the type of solvent and the temperature. An explanation of this abnormality was offered by Thiele. He postulated that only one bond is needed to hold two carbon atoms together and that in the case of a double bond only one valency and part of the other valency is used, the other part of the second valency being 'free'. Thus if we represent these partial valencies with dotted lines butadiene may be written

$$CH_2{\ldots}CH{-}CH{\ldots}CH_2$$

The two middle valencies satisfy each other and the actual state of butadiene is

$$CH_2{\ldots}CH{\ldots}CH{\ldots}CH_2$$

Thus 1 : 4 addition becomes normal and to be expected. The theory is so satisfactory, however, that it is difficult to account for 1 : 2 addition. The modern explanation of this abnormal addition depends upon the electromeric effect on butadiene.

Thomsen-Berthelot Principle
J. Thomsen (1854) and M. Berthelot (1867) suggested that the heat evolved in a chemical reaction was a measure of the 'affinity' of the reacting substances. This view, which implies that only exothermic reactions can occur spontaneously, must be incorrect, as for each exothermic reaction there must be an endothermic reaction.

Thomson Effect
Heat is absorbed or liberated when an electric current passes through a single homogeneous but unequally heated conductor. The coefficient

of the Kelvin effect for a given metal is defined as the heat absorbed per second when a current of one ampere flows from one point in the metal to another whose temperature is $1\,^\circ$C higher, over and above the heat developed according to **Joule's Law**. Also called the **Kelvin Effect**.

Thomson-Freundlich Equation
The solubility of component B in a solid solution of a is given by

$$\ln \frac{[B_r^a]}{[B_{r\to\infty}^a]} = \frac{2M\,\sigma}{rRT\rho}$$

where $[B_r^a]$ is the concentration of B at saturation for particles of radius r and $[B_{r\to\infty}^a]$ is the concentration for large particles, M is the atomic weight and ρ the density of component B. σ is the interfacial free energy or surface tension.

Thomson Parabolas
Thomson showed the existence of isotopes for various elements by passing ions of the elements through crossed magnetic and electric fields and on to a photographic plate which then showed up a pattern of parabolas called Thomson parabolas. Each parabola corresponds to a particular mass/charge ratio and hence to a particular isotope.

Thomson Scattering
The scattering, with no change of wavelength, of electromagnetic radiation by electrons. Thomson calculated the intensity of radiation scattered by a free electron to be

$$I = \frac{8\pi}{3} \left(\frac{e^2}{mc^2}\right)^2 I_0$$

where I_0 is the intensity of the primary radiation. I/I_0 is called the scattering coefficient for a free electron and gives an effective electron radius of e^2/mc^2. (The electrons are usually in a bound state but because the nuclei possess large mass, only acceleration of the electrons need be considered.) This type of scattering is unimportant for radiation of visible wavelengths but is important at x-ray wavelengths. *See also* **Appendix-Thomson Cross-section.**

Thomson's Formula

The period of oscillation of the current during the discharge of a capacitor is given by

$$T = \frac{2\pi}{\sqrt{(1/LC - R^2/4L^2)}}$$

where C is the value of the capacitance, and L and R are the associated series inductance and resistance.

Thomson's Theoretical Gas/Liquid Curve

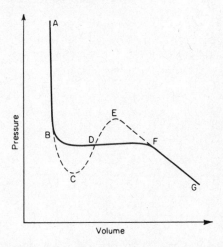

Thomson's Theoretical Gas/Liquid Curve

P/V curves for real gases follow the curve ABFG. The suggestion was made by J. J. Thomson that the ideal behaviour of a gas should be represented by the curve ABCDEFG thus emphasizing the essential continuity between the liquid and gas phase. This postulate may have no theoretical significance but it enables the P/V relation to be expressed as a simple mathematical equation whereas it is difficult to express discontinuous behaviour of the type represented by ADG.

Thomson (Thermocouple) Relations

For two conductors A and B connected at their two ends with a

299

temperature difference ΔT between the junctions, the Peltier heat is given by

$$\frac{d\Sigma}{dT} = \frac{\Pi_{AB}}{T}$$

where Σ is the Seebeck voltage between the junctions and Π_{AB} is the Peltier coefficient for a junction between metals A and B.

The Thomson heat is given by

$$\mu_A - \mu_B = -T \frac{d^2\Sigma}{dT^2}$$

where μ_A and μ_B are the Thomson coefficients for conductors A and B. (*See also* **Peltier, Seebeck** *and* **Thomson Effects**.)

Thue's Theorem
The equation

$$a_0 x^n + a_1 x^{n-1} y + \ldots + a_n y^n = C$$

where $n \geqslant 3$ and C is integral, and where the form on the left is irreducible, cannot have an infinite number of integer solutions x, y.

Tischenko Reaction
All aldehydes can be made to undergo the **Cannizzaro Reaction** in the presence of aluminium ethoxide. Under these conditions the ester is formed.

$$2CH_3CHO \xrightarrow{Al(OC_2H_5)_3} CH_3.CO.OCH_2.CH_3$$

Tollen's Aldehyde Test
Solution A: 10% solution $AgNO_3$.
Solution B: 10% solution NaOH.

These two solutions are mixed in a perfectly clean test tube and 0·880 ammonia is added drop by drop until the precipitate of silver oxide just redissolves. The addition of a dilute solution of an aldehyde produces a silver mirror in the cold. It is dangerous to boil the mixture because of the production of fulminates.

Tollen's Test (for Glycuronates in Urine)
To 5 c.c. of urine in a rather wide test tube add about 1 c.c. of a
1% solution of naphthoresorcin in alcohol and 5–6 c.c. of concentrated
hydrochloric acid. Heat slowly to boiling and boil for 1 minute shaking
the tube the whole time. Leave for 4 minutes, then cool under the tap
and shake with an equal volume of ether. The ether is coloured violet
to red and when examined spectroscopically shows two bands, one on
the D line and one to the right of it.

Tollen's Test (for Pentoses in Urine)
To 5 c.c. of urine add an equal volume of concentrated hydrochloric
acid and a little phloroglucinol, heat the mixture to boiling on a water
bath. A cherry red coloration develops and the solution shows an
absorption band between D and E. On cooling a dark precipitate
separates out which when filtered off and dissolved in alcohol shows
the same absorption band as the original mixture.

Torr *See* **Appendix**

Torricellian Vacuum
The space enclosed above a column of mercury when a sealed tube is
filled with mercury and inverted so that its lower end is inserted in a
volume of mercury. The height of the mercury column will then be
limited to that which the atmosphere can support. The enclosed space
is evacuated except for mercury vapour.

Torricelli's Law of Efflux
The velocity of efflux of a liquid through an orifice is equal to that
which a body would attain in falling freely from the free surface of the
liquid to the orifice.

Townsend Coefficient
The number of ionizing collisions per centimetre of path of a charged
particle in the direction of the applied electric field.

Townsend Discharge
A discharge in which additional ionization is solely due to ionization of
the gas by electron collisions.

Traube's Rule
Traube (1891) found that for dilute solutions, the concentration of a
member of a homologous series at which equal lowering of surface

tension was observed decreased threefold for each additional methylene group in any given series.

Trouton's Law

The molar latent heat of vaporization of a liquid divided by its boiling point at atmospheric pressure on the absolute scale of temperature is equal to approximately 23, if the latent heat is expressed in calories. This generalization was first discovered by A. Pictet (1876) and rediscovered by W. Ramsay (1877) and F. Trouton (1884).

Tschebyscheff *See* Chebyshev

Twaddle (Twaddell) Scale

A scale of specific gravity, where degrees Twaddle are given.

$$°Tw = (sp.gr. - 1) \times 200$$

Tyndall Effect

If a strong beam of light is passed through a true solution, the path of the beam cannot be seen. If the solution, however, contains colloidal particles, the beam becomes visible when viewed at right angles to its path by virtue of the scattering of light by the particles. The ultramicroscope depends on the effect.

The intensity of the scattered light is proportional to the fourth power of the frequency; hence the sky appears blue because of the scattering of sunlight by small dust particles. (*See also* **Rayleigh Scattering.**)

U

Uffelmann's Reaction (for Lactic Acid)

When a few c.c. of Uffelmann's reagent are treated with a solution of lactic acid (0·4%) the violet coloration is turned immediately yellow. This reaction is not specific, other acids, such as tartaric, oxalic and citric acids, give it. Uffelmann's reagent is prepared by treating a solution of phenol with very dilute ferric chloride until it becomes amethyst-violet in colour.

Ullmann Reaction

Triphenylamine may be prepared by heating together diphenyl-amine, iodobenzene, potassium carbonate and a little copper powder in nitrobenzene.

$$2(C_6H_5)_2NH + 2C_6H_5I + K_2CO_3 \xrightarrow{\text{Cu}} 2(C_6H_5)_3N + CO_2 + 2KI + H_2O$$

Urbach's Rule

Optical absorption in many semiconductors increases exponentially with energy near the absorption edge with a proportionality coefficient in the exponent of σ/kT where σ is a constant which is approximately 1.

V

Van Allen Radiation Belts
Two toroidal-shaped regions surrounding the earth with their axes coinciding with the earth's geometric axis and containing circulating particles of low energy but high intensity. Particles in the outer belt probably originate from the sun, those in the inner belt from the radioactive decay of neutrons liberated in the atmosphere by cosmic radiation.

Van Arkel and De Boer's Process
A filament growth process for the production of pure Zirconium, Boron, Silicon, Titanium and Hafnium among others depending on the fact that a volatile compound such as zirconium tetraiodide is decomposed thermally in a vacuum, on an incandescent wire.

Vandermonde Determinant
The determinant (or its transpose) having
  (i) unity for each element of the first row,
  (ii) an unspecified second row,
  (iii) the elements of the ith row as the $(i-1)^{th}$ power of the corresponding elements of the second row.

Vandermonde's Theorem
If m and n are any numbers whatsoever

$$(m + n)_r = m_r + C_1^r m_{r-1} n_1 + C_2^r m_{r-2} n_2 + \ldots + n_r$$

where

$$n_r = n(n - 1)(n - 2) \ldots (n - r + 1)$$

and
$$C_k^r = \frac{r!}{r - k! \; k!}.$$

Van der Waals Adsorption
Two main types of adsorption have been distinguished depending upon whether the association between the gas and solid is physical or

chemical in nature. Physical adsorption depends for cohesion between gas and solid on the so-called Van der Waals forces, which are the normal cohesive forces between any molecules. It is this type of adsorption which is known by the above name. In the other case the adsorption depends upon valence forces. Van der Waals adsorption is characterized by relatively low heats of adsorption of the order of 5 kcal mole^{-1}, that is, of the same order as the heat of vaporization of a gas. The equilibrium between gas and solid is reversible and is attained rapidly when the pressure and temperature are changed.

Van der Waals' Equation
An equation of state for a real gas, which may be expressed in the form

$$\left(p + \frac{a}{V^2} \right)(V - b) = RT.$$

a/V^2 is a measure of the attractive force of the molecules and is called the cohesive pressure, whilst b is the covolume which is equal to four times the actual volume of the molecules.

Van der Waals–London Interaction
The dipole interaction between two neutral atoms, given by

$$\Phi_r = - (C/r^6)$$

where C is a constant, r is the separation of the atoms and Φ_r is the potential energy of the interaction. (See also **Lennard-Jones Potential.**)

Van Slyke Method (for the Rate of Protein Hydrolysis)
The rate of hydrolysis of proteins is followed by the rate of evolution of nitrogen when the solution is treated with nitrous acid.

$$
\begin{array}{ccc}
\text{COOH} & & \text{COOH} \\
| & & | \\
\text{R--C--NH}_2 & \xrightarrow{\text{HONO}} & \text{R--C--OH} + \text{N}_2 \uparrow + \text{H}_2\text{O} \\
| & & | \\
\text{H} & & \text{H}
\end{array}
$$

Proline and hydroxy-proline do not react with nitrous acid.

305

Van't Hoff Equation (for Osmotic Pressure)

Van't Hoff pointed out in 1886 the analogy between gases and solutions. He deduced from Pfeffer's work on osmotic pressure an equation for the osmotic pressure of a dilute solution analogous to the **Clapeyron Equation** for gases

$$\pi V = RT$$

where π is the osmotic pressure, V is the dilution, T is the absolute temperature and R is a constant almost exactly equal in value to the ordinary gas constant.

Van't Hoff Factor

The osmotic pressure of a solution of an electrolyte cannot be expressed by a simple **Van't Hoff Equation**. In order to make allowance for the deviation Van't Hoff proposed a modification to the equation

$$\pi V = iRT$$

where i is known as the Van't Hoff Factor. If n molecules of an electrolyte are contained in a certain volume and the degree of dissociation of the molecules is a, then

$$i = a(n - 1) + 1.$$

Van't Hoff Isochore

This equation, which is of fundamental importance in chemistry, represents the variation with temperature of the equilibrium constant for a reaction involving gases in terms of the change in heat content and may be written

$$\frac{d \ln K_p}{dT} = \frac{\Delta H}{RT^2}$$

where K_p is the equilibrium constant and ΔH is the change of heat of reaction at constant pressure.

A similar equation for reaction at constant volume may be deduced.

Van't Hoff Isotherm

This equation expresses the change of free energy in terms of the temperature, the equilibrium constant and some factor expressing the concentration and the number of molecules of each reactant.

$$- \Delta G = RT \ln K_c - RT \Sigma \nu - \ln c.$$

where c is concentration in gm/litre and ν is the number of molecules of reactants.

Van't Hoff-Le Bel Theory

Van't Hoff Theory

In 1874 Van't Hoff and Le Bel gave independently a solution to the problem of optical isomerism. Van't Hoff postulated that the four bonds of a saturated carbon atom were directed to the corners of a tetrahedron. Le Bel's theory is similar in nature; he postulated only that whatever the spatial arrangement of the bonds, the compound Cabcd is asymmetrical. The bonds were not therefore definitely fixed in direction as in the Van't Hoff theory. Modern work especially x-ray and dipole moment studies have shown that the Van't Hoff theory is more likely to be correct. Thus the two structures shown in the figure represent lactic acid and as they are asymmetric, i.e. it is impossible to superimpose one on the other, they represent the dextro- and laevo-rotatory forms of this compound.

Van't Hoff Principle
If the temperature of a chemical reaction in equilibrium is raised the amount of the reactants formed by an endothermic reaction is increased whereas lowering of the temperature increases the amount of the products produced by an exothermic reaction.

Van't Hoff Principle of Superposition
The optical rotatory power of a substance containing a number of asymmetrical carbon atoms is equal to the algebraic sum of the contributions of each carbon atom taken alone. This contribution is independent of the configuration of the rest of the molecule.

307

Van Vleck Paramagnetism
In considering magnetic susceptibility as a function of temperature one limiting case is when the level splitting is $\gg kT$. Levels $\gg kT$ above the ground state make a contribution to the susceptibility which is independent of temperature. This term is known as Van Vleck paramagnetism and is in addition to the normal $1/T$ and diamagnetic terms.

Varley Effect
In an ionic solid an energetic charged particle or photon can ionize a negative ion to give it a positive charge and induce it to take up an interstitial position. (*See also the* **Wigner Effect.**)

Vegard's Law
In an alloy system, lattice spacings show a linear dependence on composition, having values between those for the pure elements. The law was originally proposed for mutually soluble pairs of ionic salts and exceptions to the rule are particularly numerous among metallic solid solutions.

Venn Diagram (or Euler Diagram)
A graphical method of illustrating **Boolean Algebra** in terms of an algebra of classes. It consists of a number of outlined areas (usually circles or other simple shapes) inside an enclosing rectangle containing the class. Each outlined area represents a variable which can be true or false; within the area it is true (represented by A, B, C, . . .) whereas outside the area but within the rectangle it is false (represented by

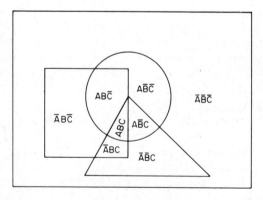

Venn Diagram

308

$\bar{A}, \bar{B}, \bar{C}, \ldots$). Where areas overlap, more than one variable is true. For instance, the enclosing rectangle could represent people. Areas enclosing A, B, C, ... would represent those people possessing particular attributes whereas those areas enclosing $\bar{A}, \bar{B}, \bar{C}, \ldots$ would represent people in whom the attributes are absent.

Verdet's Constant
The angle of rotation of the plane of polarization per unit length per unit magnetic field in the **Faraday Effect**. Verdet's Constant is approximately proportional to the square of the wavelength of the light.

Verneuil Method
A method of single crystal growth, originally intended for the preparation of artificial gem-stones, in which powder is dropped through an oxy-hydrogen flame so that it falls molten on to a crystal seed.

Villari Effect
The susceptibility of an iron wire is increased by stretching when the magnetism is below a certain value, but diminished when above that value.

Voigt Effect
When light is passed through a vapour (or liquid) in a direction which is perpendicular to an applied magnetic field, the vapour becomes doubly refracting. This arises because the absorption frequencies in the vapour are different for light polarized perpendicular and parallel to the magnetic field. The magnitude of the effect is proportional to the square of the magnetic field strength. *See also the* **Cotton-Mouton Effect**.

Volt *See* Appendix

Volta Effect
The e.m.f. established when two dissimilar metals are brought into contact with each other.

Volterra Dislocation
Consider a body such that it can be cut across without being separated,

i.e. an annulus of material. A cut can be made across one side of the ring and the cut ends can be displaced relative to each other and rejoined. The nature of this type of displacement has been considered by Volterra who showed that if the stresses in the ring are to be single-valued the surfaces of the cut must be displaced rigidly with respect to each other. The discontinuities in the components of the displacement vector along the Cartesian axes must satisfy certain conditions.

The simplest discontinuities of this kind are shown in the figure.

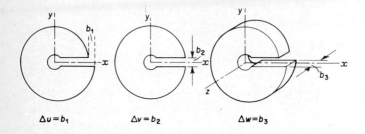

$$\Delta u = b_1 \qquad\qquad \Delta v = b_2 \qquad\qquad \Delta w = b_3$$

Volterra Equation
An integral equation of the type

$$\psi(z) = \int_a^z K(z, z_0)\, \psi(z_0)\, \mathrm{d}z_0 + \phi(z)$$

for the unknown function ψ. $K(z, z_0)$ is known as the kernel of the equation.

Von Braun Reaction
Tertiary amines displace the bromide ion from a cyanogen bromide to give a quaternary salt. An alkyl group of this compound is attacked by the bromine ion spontaneously to give a dialkyl cyanamide.

$$R_3N + BrCN \rightarrow Br^- + R_3N^+ - CN \rightarrow BrR + R_2N\!-\!CN$$

Vorlander's Rule
If the first substituent in the benzene nucleus contains an unsaturated valency then the group is m-orientating, if it is saturated then o-p substitution will take place. There are exceptions to this rule, for instance cinnamic acid (directing group CH=CH.COOH) should be meta, whereas in fact it is o-p orientating.

310

W

Wagner Rearrangement

When alcohols containing no hydrogen atoms adjacent to the alcohol group are dehydrated, dehydration and molecular rearrangement occur together.

$$(CH_3)_3.C.CH_2OH \xrightarrow{-H_2O} (CH_3)_2.C=CH.CH_3$$

Walden Inversion

The Walden inversion may be defined as the conversion of a laevorotatory optical isomer to the dextrorotatory isomer without recourse to resolution. The conversion may either be total or partial. The phenomenon was discovered when the following series of reactions were first carried out.

$$
\begin{array}{ccccc}
\text{CHOH COOH} & \xrightarrow{PCl_5} & \text{CHCl.COOH} & \xrightarrow{AgOH} & \text{CHOH.COOH} \\
| & & | & & | \\
\text{CH}_2\text{COOH} & \xleftarrow[KOH]{} & \text{CH}_2\text{COOH} & \xleftarrow[PCl_5]{} & \text{CH}_2\text{COOH} \\
\textit{l}\text{-malic acid} & & \textit{d}\text{-chlorosuccinic} & & \textit{d}\text{-malic acid} \\
& & \text{acid} & &
\end{array}
$$

Walden's Rule

According to P. Walden (1906) in an electrolyte the product $\Lambda_0.\eta_0$ is constant where Λ_0 is the equivalent conductivity at infinite dilution and η_0 is the viscosity of the solvent. The rule breaks down as the effective diameter of the ions in different solvents varies due to different degrees of solvation.

Wallach Transformation

When gently warmed with concentrated sulphuric acid, azoxy benzene rearranges to *p*-hydroxyazobenzene.

Wallis' Theorem

$$\frac{\pi}{2} = \frac{2 \cdot 2 \cdot 4 \cdot 4 \cdot 6 \cdot 6 \cdot 8 \cdot 8 \cdots}{1 \cdot 3 \cdot 3 \cdot 5 \cdot 5 \cdot 7 \cdot 7 \cdot 9 \cdots}$$

$$= \prod_{k=1,2,\ldots} \left(\frac{2k}{2k-1} \cdot \frac{2k}{2k+1} \right)$$

Wannier Exciton *See* Frenkel Exciton

Warburg's Law

The energy liberated during a complete hysteresis cycle is $\oint H\mathrm{d}M$ where H is the magnetic field strength and M the magnetization.

Waring's Formula

$$\frac{1}{x-a} = \frac{1}{x} + \frac{a}{x(x+1)} + \frac{a(a+1)}{x(x+1)(x+2)} + \cdots$$

Watt *See* Appendix

Watt's Law

The latent heat of steam at any temperature of generation added to the sensible heat required to raise the water from $0°C$ to the temperature is constant. Regnault, in 1847, showed that Watt's law was materially in error, and that the total heat of steam increased with the temperature of generation.

Weber *See* Appendix

Weber-Fechner Law

The intensity of a sensation is proportional to the logarithm of the physical stimulus which produces it. Seeing, hearing and photographic tone-reproduction follow the law closely.

Weber's Bessel Functions of the Second Kind *See* Bessel Functions

Weber's Number
A dimensionless number for flow of fluid past a body given by

$$W = \rho \, v^2 \, l/\gamma$$

where ρ is the density of the fluid, v the velocity, γ the surface tension and l a characteristic linear dimension of the body.

Weddle's Rule
A method of computing definite integrals similar to **Simpson's Rule**. It is more accurate than Simpson's Rule but n must be divisible by 6. For this rule

$$A = \frac{3h}{10} (y_0 + 5y_1 + y_2 + 6y_3 + y_4 + 5y_5 + 2y_6 + 5y_7 + y_8 + \ldots$$

$$+ 2y_{12} + \ldots + 2y_{n-6} + 5y_{n-5} + y_{n-4} + 6y_{n-3}$$

$$+ y_{n-2} + 5y_{n-1} + y_n).$$

(*See also* **Newton-Cotes Formula**.)

Weerman Degradation
An a hydroxy or a methoxy amide may be degraded to an aldehyde containing one less carbon atom by the action of a cold aqueous solution of sodium hypochlorite. This reaction has been used by Haworth to descend the sugar series.

$$\begin{array}{c} CONH_2 \\ | \\ CHOH \\ | \\ R \end{array} \xrightarrow[\text{NaOCl}]{\text{NaOH}} \left[\begin{array}{c} NCO \\ | \\ CHOH \\ | \\ R \end{array} \right] \rightarrow RCHO + Na\,NCO$$

Weierstrass' Approximation Theorem
A continuous function may be approximated to any degree of accuracy over a closed interval by a polynomial.

Weierstrass Function
A kind of elliptic function with a double pole and zero in each periodic lattice.

Thus, if

$$z = \int_{\mathscr{P}(z)}^{\infty} \frac{dt}{(g_1 t^3 - g_2 t - g_3)^{\frac{1}{2}}}$$

then $\mathscr{P}(z)$ is the Weierstrassian.

Weierstrass Inequalities

If $a_1, a_2, \ldots$ are positive numbers less than 1 whose sum is denoted by S_n then

$$1 - S_n < (1 - a_1)(1 - a_2) \ldots (1 - a_n) < \frac{1}{1 + S_n}$$

$$1 + S_n < (1 + a_1)(1 + a_2) \ldots (1 + a_n) < \frac{1}{1 - S_n}$$

where, in the last inequality it is supposed that $S_n < 1$.

Weierstrass' Infinite Product

$$\frac{1}{\Gamma(x)} = x \, e^{\gamma x} \prod_{s=1}^{\infty} \left(1 + \frac{x}{s}\right) e^{-x/s}$$

where γ denotes **Euler's Constant**.

Weierstrass' Test for Convergence

If an infinite series $a_0(x) + a_1(x) + a_2(x) \ldots$ of real or complex functions converges uniformly and absolutely on every set of values x such that $a_n(x) \leqslant M_n$ for all n, where $M_0 + M_1 + M_2 \ldots$ is a convergent comparison series of real positive terms, then the series converges uniformly.

Weissenberg Effect

The tendency of a fluid to creep up along a rotating shaft instead of being thrown out by centrifugal force.

Weisskopf Unit *See* **Appendix**

Weiss Zone Law

If any face with **Miller Indices** hkl lies in the zone $[UVW]$ defined by faces $h_1k_1l_1$, $h_2k_2l_2$ so that $U = k_1l_2 - l_1k_2$, $V = l_1h_2 - h_1l_2$ and $W = h_1k_2 - k_1h_2$ then

$$Uh + Vk + Wl = 0.$$

A zone is a set of faces with mutually parallel intersections, and $[UVW]$ is the zone symbol specifying the common direction of these intersections.

Weizmann Process
When starch or molasses is fermented with clostridium acetobylian, acetone and n-butanol are formed.

Weizsäcker's Formula
A semi-empirical expression for the binding energy of nuclei;

$$E(Z, A) = a_1A - a_2A^{2/3} - a_3\frac{Z^2}{A^{1/3}} - \frac{a_4}{A}\left(Z - \frac{A}{2}\right)^2 + \begin{cases} +a_5A^{-\frac{1}{2}} \ Z \text{ even}, N \text{ ev} \\ 0 \ Z \text{ even}, N \text{ odd or} \\ \ Z \text{ odd}, N \text{ even} \\ -a_5A^{-\frac{1}{2}} Z \text{ odd}, N \text{ od}\end{cases}$$
$$ (1) \quad (2) \qquad (3) \qquad (4) \qquad (5)$$

where $E(Z, A)$ is the binding energy, A the atomic number, Z the number of protons and N the number of neutrons ($A = Z + N$). The terms represent:

(1) A binding energy proportional to the volume.
(2) A correction to the volume term for surface effects.
(3) A correction for the Coulombic energy of the nucleus of assumed radius $A^{\frac{1}{3}}$.
(4) and (5) Terms taking into account symmetry properties in the nuclear state giving a tendency for $Z = N$ and for even Z and even N.

Weldon Process
This process, the action of manganese dioxide on hydrochloric acid, is still used to a small extent on the industrial scale especially in Spain and other countries where pyrolusite is found. Hydrochloric acid is placed in tanks constructed of granite and manganese dioxide in the form of a slurry is allowed to run in. The chlorine comes off and is carried away by pipes. The reaction is completed by blowing in steam.

The manganese chlorides formed are mixed with lime in tall iron towers and air is blown through for several hours until the manganese is re-oxidized to manganese dioxide which is then allowed to settle and re-used in the first stage.

Wentzel, Kramers, Brillouin, Jeffreys Method (W.K.B.J. Method)

A short wavelength approximation used in potential and wave-motion problems.

The solution of

$$\frac{d^2\psi_n}{dx^2} + \{k^2 - U(x)\}\psi = 0$$

can be written as

$$\psi = \exp\{\phi(x)\}$$

where $d\phi/dx$ satisfies the **Riccati's equation**

$$\left(\frac{d\phi}{dx}\right)^2 + \frac{d^2\phi}{dx^2} + q^2 = 0$$

if $q^2 = k^2 - U(x)$ is a slowly varying function of x.

Werner's Coordination Theory

A theory which accounts for the formation of stable compounds such as the chloroplatinates from platinic chloride and hydrochloric acid both of which have no residual valency in the usually accepted sense of the word. The atoms or radicals in the nucleus are said to be coordinated with the central atom and since they are not ionized must be attached by co-valencies. Certain compounds of this type exhibit isomerism, e.g. structural isomerism leading in certain cases to optical activity, etc.

Weyl's Test

A test for creatinine in urine in which the urine is mixed with alkali and sodium nitro-prusside and yields a red coloration which turns yellow. If acetic acid is added to the yellow solution, then according to Salowski, on heating, the solution becomes first green and finally blue.

Whiddington's Law

When a solid is bombarded by primary electrons to produce secondary emission (i.e. the emission of additional lower energy electrons), the energy loss of the primary electrons per unit path length is given by

$$-\frac{d\,E_p\,(x)}{dx} = \frac{A}{E_p\,(x)}$$

where E_p is the energy of the primaries and A is a constant which is characteristic of the solid. This leads to the result that the range of the primaries is proportional to the square of their energy.

Whitmore Mechanism (for the Pinacol-Pinacoline Rearrangement)
The mechanism of the Pinacol-Pinacoline rearrangement is uncertain but Whitmore suggested an explanation. Consider a compound AB–X where X is strongly electron attracting and A and B are neither electron attracting nor electron repelling. When the molecule undergoes a reaction in which X is removed, it takes its full octet of electrons with it and leaves B with an open sextet.

$$AB–X \rightarrow AB^+ + X^-$$

Then, either normal substitution of a negative ion is possible

$$AB^+ + Y^- \rightarrow ABY$$

or, if one of the atoms attached to A is a hydrogen atom, a proton may be lost and the group stabilized by olefin formation

$$H–AB^+ \rightarrow A\,^-B^+ + H^+$$

e.g. formation of ethylene from ethanol and sulphuric acid.
   If, however, B has a greater electron affinity than A, rearrangement may occur.

$$AB^+ \rightleftharpoons\,^+AB$$

This radical can now recombine with X^- or can combine with another

317

radical to give a rearranged product. The mechanism, in the case of the pinacol-pinacoline rearrangement, is

Whitworth's Theorem

If there are N events and r possible conditions, such that every single condition is satisfied in N_1 of the events, every two of the conditions are simultaneously satisfied in N_2 of the events, ... and finally all r conditions are simultaneously satisfied in N_r of the events, then the number of events free from all the conditions is

$$N - C_1^r N_1 + C_2^r N_2 - C_3^r N_3 + \ldots + (-1)^r N_r$$

where
$$C_n^r = \frac{r!}{(n-r)!\, r!}$$

Wiedemann Effect

The lower end of a vertical wire, magnetized longitudinally, twists, if free, in a certain direction when a current passes through it.

Wiedemann Franz and Lorentz's Law

The ratio of the thermal to the electrical conductivity has the same value at the same temperature for all good conductors; for different temperatures the value of the ratio is proportional to the absolute temperature.

Wiedemann's Law

The empirical law that the susceptibility χ_L of a solution containing m gm of salt of susceptibility χ_A dissolved in M gm solvent of susceptibility χ_B is

$$\chi_L = \frac{m\chi_A + M\chi_B}{M + m}.$$

Wien Effect

If a very high electric field is applied to an electrolyte so that the ion moves through many times the diameter of the ionic atmosphere during the time of relaxation, the ion should be virtually free from the retarding effect of its atmosphere and the value of the conductivity should approach that at infinite dilution. This is known as the Wien Effect (1927).

Wiener and Hopf Equation

A type of integral equation which has some importance in problems of diffraction from semi-infinite obstacles. It is of the form

$$\psi(x) = \int_0^\infty U(|x - x_0|)\, \psi(x_0) \mathrm{d}x_0 \quad x > 0$$

where $U(|x - x_0|)$ is a given real function of the absolute difference between x and x_0, $\psi(x)$ is a given function and $\psi(x_0)$ is unknown.

Wiener-Khintchine Theorem

This relates the power spectrum of a random process to the correlation function which indicates how fast the random process is changing. If the random function $y(t)$ is stationary (this means that there is no preferred origin in time for describing y) and its correlation function is

$$K(\tau) \equiv\, <y(t)\, y(t + \tau)>$$

(averaging over time t) then, by expressing $y(t)$ as a **Fourier Integral**,

$$K(\tau) = \int_{-\infty}^{+\infty} J(\omega)\, e^{i\omega\tau}\, \mathrm{d}\omega$$

319

where $J(\omega)$ is the 'spectral density' of y (the power density per unit frequency bandwidth averaged over the total time). Conversely

$$J(\omega) = \frac{1}{2\pi} \int_{-\infty}^{+\infty} K(\tau)\, e^{i\omega\tau}\, d\tau.$$

Wien's Constant *See* **Appendix**

Wien's Displacement Laws
The curve of energy of the emission of a black body at a given temperature against the wavelength of the emission passes through a maximum. By means of the following relationship Wien was able to account for the change of distribution of energy of the spectrum as the temperature was raised providing the energy curve at one temperature was known.

$$\lambda_m T = \text{constant}; \quad \frac{E_m}{T^5} = \text{constant}$$

where λ_m is the wavelength of the maximum energy E_m at absolute temperature T. *See* **Planck's Quantum Theory**.

Wigner Coefficients Alternatively called **Clebsch-Gordan Coefficients**

Wigner Effect
A neutron or charged particle with large kinetic energy can on striking an atom send it into an interstitial position thus producing a **Frenkel Defect**. To return the atoms to their original lattice positions it is necessary to anneal the material. The effect can occur, for instance, in graphite used as a moderator in nuclear reactors.

Wigner Force
A force between nucleons where there is no space or spin exchange. The force is always attractive. (*See also* **Majorana Force**.)

Wigner-Seitz Cell
A type of cell in a crystal lattice obtained geometrically by taking planes which perpendicularly bisect the lines connecting a single lattice

point to all surrounding lattice points. The minimum volume containing the lattice point that can be defined in this way is the Wigner-Seitz cell. It is a primitive unit cell as not only can all lattice space be filled by the cell, using suitable crystal translation operations, but it is also a minimum volume cell for which this is possible.

Wigner's Theorem

If ψ is an eigenfunction of H, the Hamiltonian operator (see **Hamiltonian**), describing a quantum mechanical system, and ψ corresponds to an eigenvalue E, and if R is one of a group of symmetry elements of H (for instance, R could be a rotational or reflectional symmetry operation), then $R\psi$ is an eigenfunction of H corresponding to the same eigenvalue E.

Wijs' Method

The iodine value, a measure of the unsaturation of an oil, may be determined by titration of the oil with iodine chloride in glacial acetic acid.

Willgerodt Reaction

In this reaction carbonyl compounds are converted into amides containing the same number of carbon atoms. The reaction was carried out originally by heating an aryl alkyl ketone with a solution of yellow ammonium sulphide.

$$C_6H_5COCH_3 \xrightarrow{(NH_4)_2S} C_6H_5CH_2CONH_2 + C_6H_5CH_2COONH_4$$

A modified technique is to heat the ketone with approximately equivalent amounts of sulphur and dry amine.

$$C_6H_5COCH_3 + (CH_3)_2NH + S$$
$$\downarrow$$
$$C_6H_5CH_2CH_2N(CH_3)_2 \xleftarrow{[H]} C_6H_5CH_2CSN(CH_3)_2 \xrightarrow{acid} C_6H_5CH_2COOH$$

Williamson Synthesis

This is a method of producing ethers, particularly mixed ethers. Sodium or potassium ethoxide is heated with an alkyl halide.

$$RONa + R'X \rightarrow ROR' + NaX$$

321

Williot Diagram
A graphical method of obtaining deflections at points in a structural framework under load.

Wilson's Theorem
If, and only if, p is a prime, then $(p - 1)! + 1$ is divisible by p.

Winslow Effect *See* **Johnsen-Rahbek Effect**

Witt Colour Theory
Witt pointed out that certain unsaturated or multiple bond groups were present in organic compounds which exhibited colour. Witt named them chromophores and the compound containing the chromophoric group a chromogen. The depth of colour usually increases with the number of chromophores. Thus a $-C=C-$ group produces no colour whereas $-(CH=CH)_6-$ is yellow. Certain groups deepen the colour without themselves being chromophores. These groups he called auxochromes. They are usually phenolic or basic in character the most important being $-OH$, $-NH_2$, $-NHR$, $-NR_2$. Radicals which bring about deepening of colour are known as bathochromic whilst radicals which have the opposite effect are hypsochromic.

Wohl Degradation
The aldo-sugar series may be descended by this method. For example an aldo-hexose is converted into its oxine. The reaction of acetic anhydride with this compound dehydrates the oxine group to the cyanogen group and acetylates the hydroxy groups. When the acetyl derivative is warmed with ammoniacal silver nitrate the acetyl groups are removed by hydrolysis and a molecule of hydrogen cyanide is eliminated with the formation of the pentose.

Wohlgemuth's Method (for Diastase in Urine)
Varying amounts of urine are added to a given amount of 1% soluble starch and the mixture digested for 30 min at $38°C$. After cooling a drop of dilute iodine is added. The tubes that contain a considerable amount of urine have digested all the diastase and hence no colour is obtained on the addition of starch. The experiment is repeated, if necessary, until the smallest quantity of urine which will digest all the starch is found. The diastic power D is calculated. The number of c.c. of urine needed to digest a known volume of starch is converted to the number of c.c. of urine needed to digest 5 c.c. of soluble starch.

Wölffenstein-Böters Reaction
When benzene is subjected to the action of mercuric nitrate in nitric acid, that is oxynitration; 2:4 dinitrophenol and picric acid are formed.

Wolff-Kishner Reduction
When the hydrazone or semicarbazone of an aldehyde or ketone is heated with potassium hydroxide or sodium ethoxide in a sealed tube the corresponding hydrocarbon is obtained.

$$RR'CO \xrightarrow{(NH_2)_2} RR'C{=}NNH_2 \xrightarrow[\text{pressure}]{KOH} RR'CH_2$$

Wronskian
Given a second order ordinary differential equation, its any two solutions, say y_1 and y_2, are said to be independent, i.e. not proportional to each other, when their Wronskian

$$\Delta(y_1, y_2)$$

does not vanish identically.

$$\Delta(y_1, y_2) = y_1 \frac{dy_2}{dx} - y_2 \frac{dy_1}{dx}$$

Wulff Theorem
The surface tension of a solid-fluid interface varies with the crystallographic direction of the solid surface. This variation may be represented by a surface, the **Wulff surface**, which is obtained by taking the termini of radial vectors of length proportional to the surface tension for the particular crystallographic direction. If planes are drawn perpendicular to the ends of these vectors, then the shape of the body lying inside these planes is, by Wulff's theorem, the shape of the crystal at equilibrium in the fluid. If there are pronounced cusps in the Wulff surface, then plane faces will be present on the crystal.

Würtz-Fittig Reaction
Homologues of benzene may be prepared by warming an ethereal solution of alkyl and aryl halides with sodium.

$$C_6H_5Br + Br\,C_2H_5 \xrightarrow{Na} C_6H_5C_2H_5 + 2NaBr$$

Obviously diphenyl and n-butane are formed at the same time.

323

Würtz Reaction

When an ethereal solution of an alkyl bromide or iodide is treated with sodium the hydrocarbon is formed.

$$RX + R'X + 2Na \rightarrow R - R' + 2NaX$$

It has been found that the Würtz reaction gives good yields only for paraffins containing even numbers of carbon atoms and that it usually fails with tertiary alky halides.

Y

Young's Modulus
The ratio of the stress to the fractional increase in length within the limit of proportionality for a solid subjected to a uniform tension along one axis.

Yukawa Potential
A potential function V used by Yukawa in his investigation of atomic nuclear fields. For a radially symmetric charge distribution, the potential at distance r from the centre of the charge distribution is

$$V = \frac{g}{r}\,e^{-r/L}$$

where g and L are constants. A function of this type can be obtained as a static solution of the **Klein-Gordon Equation.**

Z

Zeeman Effect

Zeeman found that when a beam of light passed through a magnetic field any spectral line of frequency $p/2\pi$ was split up into

(a) two lines of frequency $(p \pm \Delta p)/2\pi$ for the light emitted along the lines of magnetic force (Longitudinal effect),

(b) three lines of frequency $p/2\pi$, $(p \pm \Delta p)/2\pi$ for the light emitted at right angles to the lines of force (Transverse effect)

where $\Delta p = eH/2mc$. H is the magnetic field.

In (a) the two lines are circularly polarized with right-handed (for the line of lower frequency) and left-handed (for the line of higher frequency) polarization with respect to the magnetic field.

In (b) the radiation is composed of three plane-polarized constituents, the polarization of the central component being perpendicular to that of the other two and parallel to the lines of magnetic force.

This is the simple Zeeman effect, theoretically explained by Lorentz using classical electron theory. More complex resolutions observed in practice (anomalous Zeeman effect) are now understood and are found to be in agreement with the quantum theory of atomic structure and radiation. (*See also* **Appendix-Zeeman Displacement.**)

Zener Effect

At high electric fields, rectification at a semiconductor metal interface breaks down due to transportation of electrons through the insulating section by means of the tunnel effect. *Proc. Royal Soc.* **A145**, 523, 1954.

Zintl's Rule

In saltlike ionic compounds, only those elements that precede the noble gases by one to four places are able to become negative ions. Exceptions to the rule have been found in alloys of indium and gallium and the rule should be modified to include elements up to five places before a noble gas.

Appendix

A LIST OF NAMED UNITS

Amagat
A unit of volume often used in the study of the equation of state of gases. By definition the molar volume of a gas at 0°C and 1 atmosphere is 1 amagat. The exact value of the unit depends upon the gas considered but is approximately 2.24×10^4 cm^{3}mole^{-1}.

Ampere (A or amp)
The ampere is that constant current which when maintained in two parallel conductors of infinite length and negligible cross-section placed one metre apart in a vacuum produces a force of 2×10^{-7} newtons per metre length. This is the absolute unit and replaces the international ampere which is that current which under specified conditions deposits 0.001118 grammes of silver from a silver nitrate coulometer. It is equal to 0·99986 absolute amperes. The abampere or c.g.s. electromagnetic unit of current is 10 absolute amperes.

Ampere-turn (At)
A unit of magnetomotive force. A coil of n turns carrying a current of A amperes gives rise to a magnetomotive force of nA ampere-turns.

Ångström Unit (Å)
A unit of wavelength first introduced into spectroscopy by A. J. Ångström, who in 1868 mapped the spectrum of the sun in terms of it. It is ten-millionths of a millimetre or 10^{-8} cm and is called either a tenth-metre or Ångström unit and given the symbol A.A.U. or Å. It is now used in physics and chemistry to express atomic or molecular dimensions.

Balmer

A term sometimes used for the unit of wave-number, i.e. it is the number of waves in a centimetre and has units of cm^{-1}. *See also* **Rydberg** *and* **Kayser**.

Bel

A measure of attenuation named after Alexander Bell. *See the* **neper**.

Bohr Magneton (μ_B)

A unit of magnetic moment.

$$\mu_B = \frac{eh}{4\pi mc} = 9 \cdot 273 \times 10^{-21} \text{ erg gauss}^{-1}.$$
$$= 9 \cdot 273 \times 10^{-24} \text{ ampere-metre}^2.$$

It represents the fundamental unit of magnetic moment per atom.

Bohr Radius (a_0)

$$a_0 = \frac{\hbar^2}{m\,e^2} = 0 \cdot 529166 \times 10^{-8} \text{ cm.}$$

Boltzmann's Constant (k)

$$k = 1 \cdot 38041 \times 10^{-23} \text{ joule deg}^{-1}$$
$$= 8 \cdot 6167 \times 10^{-5} \text{ eV deg}^{-1}.$$

Brewster (B)

The unit used to measure the stress-optical constant when a material shows birefringence under an applied stress. A brewster is the number of angströms per mm path that one component of the light is retarded relative to the other if a stress of 1 bar (atmosphere) is applied. It is equivalent to 10^{-13} cm^2 dyne^{-1}.

Brig

A ratio of two units to base 10.

Compton Wavelength (λ_C)

$$\lambda_C = \frac{h}{mc} = 24 \cdot 2621 \times 10^{-11} \text{ cm.}$$

See also **Compton Effect**.

Coulomn (C)

The quantity of charge which crosses a section of a circuit in one second when there is a constant current of one ampere. The absolute coulomb is defined from the absolute ampere and the international coulomb from the international ampere. In the c.g.s. electrostatic system, the statcoulomb is defined in terms of the electrostatic force between charges. A statcoulomb placed one centimetre away from a like charge repels it with a force of one dyne. 1 coulomb = $2 \cdot 996 \times 10^9$ statcoulombs.

Curie (c)

That quantity of any radioactive substance which has a decay rate of $3 \cdot 7 \times 10^{10}$ disintegrations per second.

Debye (D)

A unit of electric dipole moment used for the dipole moments of molecules.

$$1 \text{ debye} = 1 \times 10^{-18} \text{ e.s.u.}$$
$$= 3 \cdot 33 \times 10^{-28} \text{ coulomb-cm.}$$

Farad (F)

A capacitor has a capacitance of one farad if a charge of one coulomb raises the potential between its plates by one volt. This unit is very large and the practical unit is the micro-farad equal to 10^{-6} farad.

Faraday

The faraday is the electric charge carried by one gramme equivalent of an ion and is equal to 96,500 international coulombs or 96,490 absolute coulombs.

Fermi

A unit of length equal to 10^{-13} cm. Radii of nuclei are of this order of magnitude.

Galileo or Gal

Unit of acceleration in the c.g.s. system of units equal to 1 cm per sec^2.

329

Gauss (G)

If a straight wire is passed through a magnetic field so as to cut it at the rate of 1 cm per second perpendicular to the direction of the induction then the value of induction necessary to produce an electromotive force of 1 abvolt per cm length of the wire is one gauss. It is equal to 1 maxwell per cm^2.

Gilbert (Gb)

The c.g.s. electromagnetic unit of magnetomotive force. 1 gilbert = $10/4\pi$ ampere-turns.

Hefner

The hefner, a unit of luminous intensity, is the intensity of a lamp of specified design burning amyl acetate called the Hefner Lamp. 1 hefner = 0·9 international candle.

Henry (H)

The henry is that inductance in which an induced electromotive force of one volt is produced when the inducing current is changed at the rate of one ampere per second. In practice the millihenry is of convenient magnitude.

Hertz (Hz)

Unit of frequency equal to cycles per second.

Joule (J)

A unit of work defined as 10^7 c.g.s. units of work or ergs. The erg is the work done when a force of one dyne acts through a distance of one centimetre.

Kapp Line

A unit of magnetic flux equal to 6000 maxwells.

Kayser
The accepted name for the unit of wave-number, cm^{-1}.

Lambert (L)
The luminance of a uniform diffuser which emits a total flux of 1 lumen per sq cm. If the diffuser emits a flux of 1 lumen per sq ft, then it has a luminance of 1 **foot-lambert**.

Langley
A measure of radiant energy received per unit area over an integrated time.
$$1 \text{ langley} = 1 \text{ calorie } cm^{-2}.$$

Lorentz
The **Zeeman Displacement** ($e/4\pi mc^2$) expressed in wave-numbers.

Maxwell (Mx)
The c.g.s. electromagnetic unit of magnetic flux.
The maxwell is the amount of magnetic flux which acting on a unit magnetic pole will propel it with a force of one dyne. It can also be defined as the amount of flux passing through one square centimetre for a field of flux density of one c.g.s. unit. The unit magnetic pole is one which will exert a force of one dyne on another unit pole one centimetre distant *in vacuo*.

Neper (Np)
A unit of attenuation named after John Napier. It is defined by
$$N_{\text{nepers}} = \ln (I_1/I_2)$$
where I_1 is the input and I_2 the output current. A most used form of unit is now the decibel defined by
$$N_{\text{db}} = 10 \log (P_1/P_2)$$
where P_1 and P_2 are powers before and after attenuation.
10 decibels = 1 **bel**.

Newton (N)
A unit of force in the m.k.s. system. That force which induces in one kilogramme an acceleration of one metre per second per second.
1 newton = 10^5 dyne.

Oersted (Oe)

The c.g.s. electromagnetic unit of magnetic flux although prior to 1932 this term was used as a practical unit of magnetic reluctance (resistance). If a unit magnetic pole—a unit magnetic pole exerts a force of one dyne on another unit magnetic pole one centimetre away—is placed in a vacuum traversed by a magnetic field of one oersted, the pole experiences a force of one dyne in the direction of the field.

Ohm (Ω)

The practical unit of resistance, the ohm, is that resistance through which a difference of potential of one absolute volt will produce a current of one absolute ampere. The international ohm is the resistance offered at $0°C$ to an unvarying current by a column of mercury of mass 14·4521 grammes of constant cross sectional area and 106.300 centimetres in length. This quantity is sometimes called a legal ohm. The unit of conductivity is the reciprocal ohm or mho.

Pascal

A unit of stress.

$$1 \text{ pascal} = 10 \text{ dynes cm}^{-2}$$
$$= 1 \text{ newton m}^{-2}.$$

Planck's Constant (h)

$$h = 6·6249 \times 10^{-27} \text{ erg sec.}$$
$$= 6·6249 \times 10^{-34} \text{ joule sec.}$$

Poise (p)

A unit of dynamic viscosity. If the shearing stress per unit area, τ, in a liquid is expressed in dynes and the velocity gradient perpendicular to the direction of flow is du/dz then the viscosity η defined by $\eta = \tau/(du/dz)$ is in units of poise (or alternatively gramme per cm per sec).

Röntgen (r)

A measure of the ionizing effect of γ or x-rays in air. A röntgen is the quantity of radiation which produces, in 0·001293 grammes of air, ions carrying 1 e.s.u. of positive or negative charge.

332

Rutherford (rd)
That quantity of any radioactive substance which has a decay rate of 10^6 disintegrations per second.
(*See also* **Curie.**)

Rydberg
A measure of wave-number.

$$1 \text{ rydberg} = 109{,}737 \text{ cm}^{-1} \text{ (or \textbf{kaysers})}.$$

See also **Balmer.**

Sabin
A unit of absorption in acoustics. If V is the volume in cubic feet of an enclosure and T the reverberation time in seconds, the absorption a in sabins is given by

$$a = 0 \cdot 161 \; V/T.$$

Siegbahn X Unit
A unit of wavelength defined by the fact that at $18°C$ the grating constant of the cleavage planes of calcite is given by $d_{18} = 3{,}029.45$ X units. 1 Ångström $= 1{,}002.06$ X units.

Siemens
A practical unit of conductance equivalent to the mho (reciprocal ohm).

Sommerfeld Fine Structure Constant (a)

$$a = \frac{2\pi e^2}{ch} = 7.297 \times 10^{-3} \simeq \frac{1}{137}.$$

A constant appearing in the relativistic correction to the energy of the electron orbits in the atom.

Stefan's Constant (σ)

$$\sigma = 5.6694 \times 10^{-12} \text{ joule cm}^{-2} \text{ deg}^{-4}.$$

See also **Boltzmann's Law of Radiation.**

Stokes (S)
A unit of kinematic viscosity. When the dynamic viscosity η is

measured in **poise** and the density ρ is measured in grammes per cubic centimetre, the kinematic viscosity (η/ρ) is in stokes (or cm^2 per sec).

Stormer Unit (S)

The trajectories of charged particles in the field of a magnetic dipole become independent of the charge Ze, momentum p and mass m of the particles if all the lengths are expressed in Stormer units given by

$$S = \left(\frac{Zem}{pc}\right)^{\frac{1}{2}}.$$

The unit has particular application to charged nuclei entering the earth's magnetic field.

Svedberg

Rate of sedimentation of 10^{-13} cm per second for unit acceleration; i.e. it is the rate of sedimentation under a force of one dyne. The unit has application in centrifuging.

Talbot

A unit of luminous energy. A luminous flux of one talbot per second is one lumen, where a lumen is the luminous flux emitted within unit solid angle (steradian) by a point source having a unit intensity of one candle.

Tesla (T)

A unit of magnetic flux density.

$$1 \text{ tesla} = 1 \text{ weber m}^{-2} = 10^4 \text{ gauss.}$$

Thomson Cross-section

$$\frac{8\pi r_0^2}{3} = 0.66516 \times 10^{-24} \text{ cm}^2.$$

r_0 is the effective electron radius given by e^2/mc^2 (*see* **Thomson Scattering**).

Torr

Named after Torricelli, it is a pressure of 1013250/760 dynes cm^{-2}. A standard atmosphere has a pressure of 760 torr.

Volt (V)

The volt is the electric potential or electromotive force between two points in a conductor carrying a constant current of one ampere when one watt of power is required to maintain the current. The absolute volt is defined from the absolute ampere and the international volt from the international ampere. The abvolt is the c.g.s. electromagnetic unit and it is the potential between two points when one erg of work is required for one abcoulomb to move from one point to the other. 1 abvolt = 10^{-8} volt.

Watt (W)

The unit of electrical power. It is represented, practically, by the work done at the rate of one joule or 10^7 ergs per second. It is also equivalent to the product of volts and amperes.

Weber (Wb)

The accepted unit of magnetic flux is the volt-second or weber which is equal to 10^8 maxwells. A change of flux at a uniform rate of one weber per second in a coil of N turns induces an e.m.f. of N volts.

Weisskopf

A unit used to express the transition probability of nuclei from one state to another.

Wien's Constant

$$\lambda_m T = \text{constant} = 0.28978 \text{ cm deg.}$$

(*See also* **Wien's Displacement Laws.**)

Zeeman Displacement

The displacement in wave-numbers for unit magnetic field;

$$\frac{e}{4\pi m c^2} = 4.66879 \times 10^{-5} \text{ cm}^{-1} \text{ gauss}^{-1}.$$

(*See also* **Zeeman Effect.**)